OECD Studies on Water

Meeting the Challenge of Financing Water and Sanitation

TOOLS AND APPROACHES

This work is published on the responsibility of the Secretary-General of the OECD. The opinions expressed and arguments employed herein do not necessarily reflect the official views of the Organisation or of the governments of its member countries.

This document and any map included herein are without prejudice to the status of or sovereignty over any territory, to the delimitation of international frontiers and boundaries and to the name of any territory, city or area.

Please cite this publication as:
OECD (2011), *Meeting the Challenge of Financing Water and Sanitation: Tools and Approaches,* OECD Studies on Water, OECD Publishing.
http://dx.doi.org/10.1787/9789264120525-en

ISBN 978-92-64-12051-8 (print)
ISBN 978-92-64-12052-5 (PDF)

Series: OECD Studies on Water
ISSN 2224-5073 (print)
ISSN 2224-5081 (online)

Photo credits:
Cover © From left to right © Taro Yamada/Corbis, © iStockphoto/Roger Whiteway, © iStockphoto/Carmen Martínez Banús, © iStockphoto/Mark Tenniswood

Corrigenda to OECD publications may be found on line at: *www.oecd.org/publishing/corrigenda*.
© OECD 2011

You can copy, download or print OECD content for your own use, and you can include excerpts from OECD publications, databases and multimedia products in your own documents, presentations, blogs, websites and teaching materials, provided that suitable acknowledgment of OECD as source and copyright owner is given. All requests for public or commercial use and translation rights should be submitted to *rights@oecd.org* Requests for permission to photocopy portions of this material for public or commercial use shall be addressed directly to the Copyright Clearance Center (CCC) at *info@copyright.com* or the Centre français d'exploitation du droit de copie (CFC) at *contact@cfcopies.com*.

Foreword

Almost 900 million people cannot get clean drinking water and 2.5 billion lack access to basic sanitation. Polluted water and poor sanitation cause 1.5 million preventable child deaths per year which makes them among the biggest causes of infant mortality along with malaria and malnutrition.

This book "Meeting the Challenge of Financing Water and Sanitation" presents strategies on how finance for essential water and sanitation services can be mobilised. And it offers a set of concrete policy tools that governments can use to support these efforts.

Improving water and sanitation infrastructure will require a significant scale-up in funding in both developed and developing countries. For example, it is estimated that the US will have to invest USD 23 billion annually for the next 20 years to maintain water infrastructure at current service levels, while meeting health and environmental standards. Meanwhile, in developing countries, current spending will need to double – to about USD 18 billion per year, to expand water services and achieve the water and sanitation Millennium Development Goals. In addition, investment will be needed to maintain the existing water infrastructure in these countries, which will add another USD 54 billion spending per year.

The benefits of improved water and sanitation are massive. One dollar of investment in water and sanitation saves 4 to 12 dollars in avoided health care costs alone. African governments and G8 leaders have both recognised the challenges and committed themselves to supporting a more strategic approach as well as to the mobilisation of more financial resources.

Closing the significant gap between the funding that is currently available and the investment that is needed will require significant efforts by governments and the private sector around the world. In this context "Meeting the Challenge of Financing Water and Sanitation" provides support for these policy efforts based on ongoing OECD work and cross-country evidence from the experiences of both developing and developed countries.

One of the most under-utilised opportunities for reducing the funding gap comes from improving the efficiency of the water and sanitation sector.

This report shows how governments can implement the necessary reforms, and establish more sustainable financing for the sector relying on three basic sources of revenue - the 3Ts (*i.e.* taxes, tariffs and transfers). It also highlights how countries can mobilise repayable finance, including through innovative mechanisms such as grouped financing vehicles and microfinance.

Finally, there is an urgent need for governments to think more strategically about the water sector. Strategic financial planning assists governments to set realistic infrastructure targets that can be achieved with available resources and that are agreed in a multi-stakeholder policy dialogue. Heeding this message will go a long way towards ensuring adequate financing for the water sector and will improve the lives of millions of people around the world. The OECD stands ready to help!

Angel Gurría
OECD Secretary-General

Acknowledgments

This report was prepared by Sophie Trémolet (Trémolet Consulting, UK) with inputs from Peter Börkey, Céline Kauffmann and Alexandre Martoussevitch from the OECD secretariat in Paris. It draws on a number of recent OECD publications, including *Managing Water for All: An OECD Perspective on Pricing and Financing* (2009), *Strategic Financial Planning for Water Supply and Sanitation* (2009), *Private Sector Participation in Water Infrastructure. OECD Checklist for Public Action* (2009), *Pricing water resources and water and sanitation services* (2010a), *Innovative Finance Mechanisms for the Water Sector* (2010b) and *Benefits of Investing in Water and Sanitation: an OECD Perspective* (2011).

Several experts from the OECD secretariat contributed inputs into Part II of the report, including Valérie Gaveau (Development Cooperation Directorate), Céline Kauffmann (Directorate for Financial and Enterprise Affairs), Tatiana Efimova, Nelly Petkova and Alexander Martoussevitch (all Environment Directorate). Jim Winpenny (Wynchwood Consulting) contributed inputs into the design of the work, while Diane Binder (Trémolet Consulting) carried out initial research as well as editing for Part II. Finally, the authors are grateful to Peregrine Swann (WHO) for his comments and review.

The report was discussed and declassified by the OECD Working Party on Biodiversity, Water and Ecosystems at its meeting in March 2011.

Table of contents

Abbreviations and acronyms 11

Executive summary 13

Introduction 21
    Overview 21
    Structure of the report 22

Part I. Key issues with respect to financing water and sanitation

Chapter 1. **What are the benefits of investing in WSS?** 27
    1.1. Identifying necessary investments in WSS 28
    1.2. Estimating the benefits of investing in WSS 31

Chapter 2. **Current status of WSS and investment needs** 41
    2.1. Current status and investment needs in OECD countries and transition economies 42
    2.2. Overview of investment needs in developing countries: Reaching the MDGs 45

Chapter 3. **Where is the money going to come from?** 51
    3.1. Reducing costs and improving efficiency 53
    3.2. Closing the gap: A combination of the 3Ts 56
    3.3. Bridging the gap: Tapping repayable sources of funding 64
    3.4. Mobilising the private sector 73
    3.5. Using strategic financial planning 79

Part II. A toolbox to support effective water and sanitation policies

Chapter 4. Introduction to the toolbox 89

Chapter 5. Strategic Financial Planning for WSS at national or regional level – the FEASIBLE tool 93

    5.1. Background and rationale 94
    5.2. Description of the Strategic Financial Plan process and the FEASIBLE tool .. 95
    5.3. Where has it been applied? 98
    5.4. Lessons learned and the way forward 98
    5.5. How to get started? 101

Chapter 6. Financial planning tool for water utilities 103

    6.1. Background and rationale for developing the tool 104
    6.2. Description of the tool 105
    6.3. Where has it been applied? 106
    6.4. Lessons learned and the way forward 107
    6.5. How to get started 108

Chapter 7. Multi-year investment planning tool for municipalities 109

    7.1. Background and rationale for developing the tool 110
    7.2. Description of the tool 110
    7.3. Where has it been applied? 112
    7.4. Lessons learned and the way forward 112
    7.5. How to get started 112

Chapter 8. Guidelines for performance-based contracts 115

    8.1. Background and rationale 116
    8.2. Description of the tool 116
    8.3. Where has it been applied? 118
    8.4. Lessons learned and the way forward 120
    8.5. How to get started 120

Chapter 9. Water Utility Performance Indicators (IBNET) 123

    9.1. Background and rationale 124
    9.2. Description of the tool 125
    9.3. Where has it been applied? 126
    9.4. Lessons learned and the way forward 126
    9.5. How to get started 127

Chapter 10. **Private sector participation in water infrastructure –
  checklist for public action** 129

  10.1. Background and rationale for developing the tool 130
  10.2. Description of the tool 130
  10.3. Where has it been applied? 131
  10.4. Lessons learned and the way forward 133
  10.5. How to get started 134

References 137

Figures

Figure 1.1 The value chain of sustainable water and sanitation services 29
Figure 1.2 Impacts of water and sanitation investments on mortality
           in Marseille (France) 32
Figure 1.3 The water and sanitation benefits curve 36
Figure 3.1 Sources of finance for WSS 52
Figure 3.2 Volume and share of aid to water and sanitation 62
Figure 3.3 Evolution of investment in public private partnerships projects
           in developing countries, 1991-2009 77
Figure 3.4 Annual cash flow needs and available financial resources in
           Moldova's water supply and sanitation sector (2006) 80
Figure 5.1 Structure of the FEASIBLE tool 97
Figure 5.2 Expenditure needs versus collected user charges in Armenia
           (million dram) 98
Figure 6.1 Overview of the FPTWU model 105
Figure 6.2 Architecture of the FPTWU model 106
Figure 7.1 Steps of the Multi-Year Investment Planning Process 113
Figure 9.1 Overview of IBNET data sets 126

Tables

Table 1.1 Overall benefits of meeting the MDGs in water and sanitation 33
Table 2.1 Forecast operating and capital spending in countries covered,
          2010-29 (USD billion) 44
Table 3.1 Examples of innovative financial mechanisms in the water sector ... 69
Table 3.2 Typology of contractual arrangements between government (G) and
          the private sector (P) 76
Table 3.3 Funding gap (USD million per year) 82

Boxes

Box 1.1	Water catchment protection in New York (United States)	36
Box 2.1	The human right to safe and clean drinking water and sanitation	47
Box 3.1	Evaluating the financing gap in Sub-Saharan Africa: The Africa Infrastructure Country Diagnostic	81
Box 6.1	Implementation of FPTWU at Bishkek Water Company (the Kyrgyz Republic)	107
Box 7.1	MYIP implementation for the City Lutsk in Ukraine	113
Box 8.1	Lease contract for Yerevan Djur, Armenia	118
Box 8.2	Management contract for the Armenia Water and Wastewater Company, Armenia	119

Abbreviations and acronyms

3Ts	Tariffs, Taxes, Transfers
ACP	Africa, the Caribbean and the Pacific
BRIC	Brazil, Russia, India and China
BCR	Benefit-Cost Ratio
CBA	Cost-Benefit Analysis
CEA	Cost-Effectiveness Analysis
CRS	Creditor Reporting System
DAC	Development Assistance Committee (OECD)
DALY	Disability-Adjusted Life Year
DFID	Department for International Development (United Kingdom)
ECAs	Export Credit Agencies
EECCA	Eastern European, Caucasus and Central Asia region
ESI	Economics of Sanitation Initiative
EU	European Union
EUWI	European Union Water Initiative
FCR	Full Cost Recovery
GLAAS	Global Assessment of Sanitation and Drinking-Water
GPOBA	Global Partnership for Output-Based Aid
IBNET	International Benchmarking Network for Water and Sanitation Utilities
IWRM	Integrated Water Resources Management
JBIC	Japan Bank for International Cooperation

JICA	Japan International Cooperation Agency
JMP	Joint Monitoring Programme (WHO-UNICEF)
MDGs	Millennium Development Goals
NGO	Non-Governmental Organisation
NRW	Non-revenue water
O&M	Operation and Maintenance
ODA	Official Development Assistance
OECD	Organisation for Economic Co-operation and Development
PPIAF	Public Private Infrastructure Advisory Facility
PPP	Public-Private Partnerships
PSP	Private Sector Participation
SSIP	Small Scale Independent Provider
USD	United States Dollars
WFD	Water Framework Directive
WHO	World Health Organization
WSP	Water and Sanitation Program
WSS	Water and Sanitation Services
WTP	Willingness To Pay

Executive summary

The investments needed to deliver sustainable water and sanitation services, including the funds that are needed to operate and maintain the infrastructure, expand coverage and upgrade service delivery to meet current social and environmental expectations, are huge. Benefits from such investments for society as a whole are equally substantial. Yet, most systems are underfunded with dire consequences for water and sanitation users, especially the poorest. Providing sustainable drinking water supply and sanitation services (WSS) requires a sound financial basis and strategic financial planning to ensure that existing and future financial resources are commensurate with investment needs as well as the costs of operating and maintaining services.

WSS generate substantial benefits for the economy

Water and sanitation services (WSS) generate substantial benefits for human health, the economy as a whole and the environment. Access to clean drinking water and sanitation reduces health risks and frees up time for education and other productive activities, as well as increases the productivity of the labour force. Safe wastewater disposal helps to improve the quality of surface waters with benefits for the environment (*e.g.* functioning of ecosystems, biodiversity), as well as for economic sectors that depend on water as a resource (*e.g.* fishing, agriculture, tourism).

Such benefits usually outstrip the costs of service provision and provide a strong basis for investing in the sector. For example, in developing countries, WHO has estimated that almost 10% of the global burden of disease could be prevented through water, sanitation and hygiene interventions. Health benefits are only a small portion of overall benefits, however. WHO estimated that meeting the water and sanitation Millennium Development Goals (MDGs) could generate about USD 84 billion per year in benefits, with a benefit to cost ratio of 7 to 1. Of those benefits, three quarters would stem from time gains, the rest being driven by reductions in water-related diseases.

For such benefits to be generated sustainably, investments in a whole range of services alongside the WSS value chain need to be carried out.

Providing access to services is usually considered as a main entry point (as reflected in the MDGs) but a whole range of other investments need to be carried out in order for access to be provided in a sustainable manner. These range from protecting freshwater resources to building storage capacity or water transport networks, all the way to investments in safe disposal, treatment or re-use of wastewater. Once built, the infrastructure needs to be adequately maintained and operated, with components renewed in a timely manner, so as to provide sustainable, affordable and reliable access to water and sanitation services.

In most countries where the "access gap" is still large, providing access to water services could deliver substantial benefits, particularly if combined with sanitation and hygiene education. The cost-effectiveness of such investments is high, especially for lower-cost investments such as hygiene promotion or on-site sanitation.

In countries where "access" is no longer the most important issue, investments in WSS are also essential in order to ensure that benefits from existing infrastructure continue to be generated as well as to meet a number of environmental objectives. In many EECCA countries, for example, a sharp deterioration in service levels implies that "having a water tap does not necessarily mean having sustainable access to safe drinking water". Cross-contamination between water and sewerage networks, due to high levels of leakage, can have serious effects on public health. In OECD, benefits from generalising wastewater treatment can be substantial, although there is some evidence of diminishing returns beyond a certain point when increasing wastewater treatment standards.

Deriving global estimates of such benefits, although potentially useful from a global policy perspective, is complicated by the fact that the magnitude of these benefits can be highly dependent on local conditions and investment sequencing, among other factors. If access to water is provided without corresponding investments in sanitation, for example, this can generate temporary disbenefits, as abundant water supply can create pools of stagnant waters mixing with excreta and other types of waste (such as grey waters). Sanitation without adequate wastewater treatment can also generate disbenefits if it transforms diffuse pollution into point-source pollution.

Investments needed to generate large benefits in both OECD and developing economies

Substantial investments are needed in order to deliver expected benefits from WSS. Key challenges include the need to expand access to water and wastewater services (particularly in developing countries but also in some OECD countries), invest in replacing and maintaining ageing infrastructure and address water security and environmental concerns. Throughout the world, the challenges of providing access to safe water and sanitation are further accentuated by increasing demands from other water uses due to a variety of factors, such as population increase, agricultural water needs for food production, rapid urbanisation, degradation of water quality, and increasing uncertainty about water availability, potentially exacerbated due to climate change. Addressing these challenges will require both large capital investments for new infrastructure, ongoing investments in maintenance, repair, upgrading and operation of existing facilities.

Despite a high initial asset base, developed countries confront huge costs of modernising and upgrading their systems. The global capital costs of maintaining and developing WSS infrastructure in OECD countries plus the BRICs has been estimated at between 0.35% to 1.2% of their GDP. This corresponds to total projected annual needs of around USD 780 billion by 2015 and USD 1 037 billion by 2025, up from a current estimated expenditure on water infrastructure of USD 576 billion annually.

In transition economies, the need for maintaining and upgrading existing infrastructure is combined with sometimes significant needs to expand coverage and address the challenges of poor governance, institutional inefficiency and the deterioration of the asset base.

In developing countries, extending access should remain a key priority. There is a broad range of estimates for the costs to reach the MDGs, depending on the assumptions used on the types of investment made. According to the GLAAS report (UN-Water, 2010), the global cost estimates for meeting the drinking water and sanitation MDG target range from USD 6.7 billion to USD 75 billion per year, *i.e.* USD 33.5 billion to USD 375 billion by 2015. Current financing allocations will not be sufficient to meet the MDGs. According to OECD (2009a), roughly a doubling of the annual rate of investment is needed.

Tariffs are a preferred funding source, but public budgets and ODA will also have a role to play

Closing the financing gap will require countries to mobilise financing from a variety of sources, which may include reducing costs (via efficiency gains or the choice of cheaper service options), increasing the basic sources of finance that can fill the financing gap, *i.e.* tariffs, taxes and transfers (commonly referred to as the "3Ts") and mobilising repayable finance (including loans, bonds and equity either from the market or from public sources) in order to bridge the financing gap.

Defining how these various sources of finance can be combined should be done based on Sustainable Cost Recovery (SCR) principles. SCR entails securing future cash flows from a combination of the 3Ts, and using this revenue stream as the basis for attracting repayable sources of finance – loans, bonds and equity, depending on the local situation. This is a key departure from earlier concepts of Full Cost Recovery (FCR) which implied that tariffs alone should be sufficient to cover all costs. In practice, particularly in poor countries where affordability is a significant constraint, SCR implies that public spending will often be required to complement revenues from tariffs, at least for a transition period.

Each country is likely to adopt a different mix of the 3Ts to meet WSS's financing needs. Most countries have used public transfers (either from their own government or from external sources) to fund the development of WSS, particularly for capital expenditure. As countries develop and WSS systems become more mature, there tends to be a shift towards more use of commercial finance, reimbursed by growing cash flows from user charges (*i.e.* tariffs). For example, whereas tariffs represent 90% of direct financial flows to the sector in France, they only account for about 40% in Korea, 30% in Mozambique or as little as 10% in Egypt (OECD, 2009d).

The mix of the 3Ts that is adopted by each government can have a substantial impact on the efficiency of the services. For example, in the US, switching from grant financing for capital investment (as used in the 1980s) to reliance on subsidised loans with long tenures and low interest rates (from the 1990s) brought significantly improved capital investment efficiency. This underlines the importance of strategic financial planning to find the right mix of the 3Ts for achieving water and sanitation targets and leveraging repayable sources of finance (OECD, 2009a).

Any strategic financial planning (SFP) exercise should start with evaluating the potential for generating financial resources via reducing the costs and improving the efficiency of existing water systems, as inefficiencies are often responsible for important losses within the sector. The scope for making such gains is particularly high in developing countries. Choice of hardware and technologies can also make big differences to costs. For example, the per capita cost of household connections is over three times higher than the costs of a stand post in Africa and Latin America.

Tariffs can provide an important source of revenues, although the potential for raising tariffs depends on affordability constraints. Apart from a few exceptions, in OECD countries, operating costs are by and large covered by tariffs but the coverage of capital costs varies substantially. WSS tariffs represent only a small share of average household incomes in OECD countries (ranging from 0.2% in Korea to 1.2% in Poland) although these average figures hide substantial variations, with areas of significant "water poverty". In developing countries, cost covering tariffs are much less prevalent, despite the fact that there are many cases where consumers could afford to pay much more. For example, in Egypt the average user charges for WSS represent less than 1% in household expenditure. However, there are also many places where serious household affordability issues prevent further increases, unless social protection measures are being introduced (OECD, 2009a).

Public budgets still represent an important share of revenue for the WSS sector and are likely to play a significant role for the foreseeable future. This is especially true where household affordability is an important constraint. In order to be efficient and effective, however, subsidies should be predictable, transparent, targeted and ideally taper off over time. While public funds are limited by budgetary constraints and multiple demands from other sectors, there is scope for increasing public budget spending. In particular, several developing countries only allocate a small portion of their GDP to the water and sanitation sector. Among the countries that had responded, Burkina Faso was the country that spent most on water and sanitation combined as a percentage of its GDP (with an estimated 3% of GDP), while countries with the lowest expenditure on the sector as a percentage of their GDP included South Sudan, Ivory Coast but also the Philippines. In the context of the economic crisis, tax transfers are only likely to surge where stimulus packages target the water sector.

Official Development Assistance in the form of grants may be able to play a role in *closing* the financing gap in transition and developing countries, while concessional loans are a potential substitute or complement for market-based repayable finance that helps to *bridge* the financing gap. The share of ODA to water and sanitation varies across recipient countries. In some countries ODA subsidises most investments, while in others it plays a more marginal role. ODA has an important role to play both as a source of finance and of capacity development for the provision and financing of water services. It can also have a catalyzing effect by reducing bottlenecks (particularly capacity constraints), ensuring access to the poor, and harmonising and aligning assistance with national strategies. After a temporary decline in the 1990s, aid to water and sanitation has risen sharply since 2001. In 2007-08, total annual average aid commitments to water and sanitation amounted to USD 7.4 billion. As noted in OECD/WWC (2008), bilateral aid to water increased at an average annual rate of 24% over the period 2002-06 and multilateral aid also rose by 21% annually.

There are, however, issues with how ODA is currently being allocated, with some countries receiving a disproportionate share when compared to their needs, and imbalances between urban and rural areas within a particular country, for example. In times of economic crisis, ODA is likely to be increasingly needed to fill the gap and a number of international organisations have indeed seen a growing demand for their services. Given rising pressures on public finances in donor countries, however, total ODA resources for the sector are unlikely to grow significantly, which means that these scarce resources will need to be spent strategically so as to maximise their leveraging capacity and effectiveness. Areas where ODA can have a catalysing effect include supporting the financial planning process, ensuring access to services by the poor and supporting the development and use of risk-management mechanisms that can help attract private funding.

Market-based repayable finance is needed to cover high up-front capital investment costs

Private funding, referred to as "market-based repayable finance" in the report, can come in the form of debt finance (including loans from commercial banks or microfinance institutions, bonds issued through capital markets, project finance) and equity finance (from private businesses, capital markets or private equity funds). Debt financing has been the backbone of most infrastructure investment in developed countries. In developing countries, water companies traditionally rely on bank loans to finance capital investments (especially concessional loans from development finance institutions) but other forms of finance, such as bond finance, project finance or equity finance are gradually emerging with some isolated examples, usually in countries where capital markets are comparatively developed, such as in India, Brazil, the Philippines or South Africa.

Financial innovation can play a major role to increase the attractiveness of the WSS sector for market-based repayable finance, and ODA can play a catalytic role in this area. Examples of such innovation can include the blending of public and private finance or the use of public guarantees (to reduce the costs of borrowing). Given that most WSS operators tend to operate at the local level, they may face difficulties due to the lack of financing opportunities at sub-sovereign level. Such constraint can be overcome in a number of ways, including through the issuance of municipal bonds, the establishment of pooled funds or mechanisms to increase lending at the sub-sovereign level (such as guarantee funds). Other types of initiatives, such as the development of credit rating systems or the establishment of project preparation facilities, can help with increasing transparency and improving the quality of projects seeking financing, given that the "lack of good projects" is often cited as a major constraint.

The private sector, as such, is unlikely to bring significant financing without an adequate business environment. Earlier expectations that introducing private sector participation into the management of WSS companies in developing countries would help attract financial resources to the sector have materialised in some countries but not everywhere. Yet there is strong evidence that the private sector is effective at controlling costs and achieving efficiency gains, which can be a major source of savings for the sector and an important step towards financial sustainability and creditworthiness, so as to strengthen the sector's ability to mobilise repayable finance.

Strategic financial planning can help governments move forward

The extent to which each source of finance can generate additional funds will be highly location-specific and depend on the overall environment and on the willingness of governments to set realistic objectives and to adopt reforms so as to improve the efficiency and creditworthiness of existing service providers.

Governments have to set realistic objectives for the development of the WSS sector, checked against available resources, and agreed in a multi-stakeholder policy dialogue (a process termed "strategic financial planning, or SFP")". Strategic financial planning must be carried out in the context of broader sector planning that address roles and responsibilities of government agencies, policy priorities and related legislative and regulatory reforms in order to ensure that a package of measures that can realistically be financed is being put forward.

Countries where most benefits are to be reaped, *i.e.* where the access gap is the largest, are also the ones where the financing gap is the most glaring and will be most difficult to fill/bridge. Where the financing gap remains substantial, public funding (in the form of domestic government funding or ODA) could potentially play a critical role in terms of leveraging other forms of finance and in providing protection for the poor. This would be where reforms to improve the effectiveness of service delivery and lowering of capital costs would be most needed.

The water and sanitation sector must include a full range of financing approaches, making the most of potential efficiency gains, adjusting targets and combining funding from both public and private sources, in order to meet its investment needs and successfully maintain and expand service. To achieve this, policy makers and water service providers need to engage in a process of strategic financial planning so as to identify what needs to be financed, how much additional resources can be generated from existing sources and how the performance of utilities can be improved to generate such efficiency gains and mobilise external financing.

Information on some of these financing sources tends to be patchy, however, which makes it difficult to reliably evaluate the gap between needs and available funding. For example, some financial information is available for central government and external donors spending, but information on subnational and local government expenditures is seldom aggregated at a national level. In addition, because funding for sanitation and hygiene is often spread over several different institutions, budget data are less available for sanitation and hygiene than for drinking water. Data on private-sector investments (ranging from large private operators, informal providers, households or remittances) is notoriously difficult to collect, although they potentially represent an important source of funding for the sector.

To provide support to governments and water and sanitation service providers, the OECD (in conjunction with a number of other international organisations) has developed a series of tools, including financial planning tools for national and local governments (such as the FEASIBLE financial model and the Multi-year Investment Planning tool presented in chapters 5 and 7), as well as for water utilities (presented in chapter 6), benchmarking and performance tools (such as IBNET presented in chapter 9 and the Guidelines for Performance-based contracts presented in chapter 8) and a checklist for public action on private-sector participation (chapter 10). These tools have been successfully tested and used in a number of OECD and developing countries. They have proven to provide economics-based analysis and approaches capable of supporting sound policy dialogue and decision-making that moves the reform agenda forward.

Introduction

Overview

The benefits of investing in water and sanitation services are very substantial. An adequate and dependable source of water is needed to sustain human life, economic development, and the integrity of ecosystems. Investment in drinking water and sanitation services can yield substantial benefits, with benefit-cost ratios that are consistently above one. According to the JMP, around 884 million people lack access to improved water sources and 2.6 billion are without access to basic sanitation. Approximately 10% of the global burden of disease could be prevented with improvements to water, sanitation and hygiene and better water resource management worldwide. The burden of water-related diseases falls disproportionately on developing countries and particularly on children under five, with 30% of deaths of such children attributable to inadequate access to water and sanitation. Wastewater from domestic and industrial uses often reaches the environment untreated or insufficiently treated, resulting in major impacts on surface waters and associated ecosystems as well as economic activity that uses these resources.

The investments needed to generate such benefits are also enormous. According to previous OECD estimates, investment needs in the water sector dwarf investment requirements in other infrastructure sectors in developed and developing countries alike. Yet, the sector remains woefully under-funded, with a large estimated financing gap, particularly in the least developed countries where the challenge of increasing access is substantial.

The financing gap may be reduced in a number of ways, starting from a reduction in operating costs, efficiency gains and a switch to less capital-intensive investment options. The 3Ts (tariffs, taxes and transfers) will need to be increased to fill the financing gap. Repayable financing, from market sources (*i.e.* commercial loans, bonds and equity) could also be increased through adequate use of financial innovation (and in some cases, the use of public funds to leverage private financing), although some of these innovations are likely to be relevant only for a small sub-set of countries.

In the short to medium-term, therefore, strategic financial planning is required to assess what would be the best combination of measures and financing sources to finance the continuing operations, maintenance and expansion of these critical services.

This report provides a comprehensive overview of key issues with respect to financing the water and sanitation sector, and presents a number of tools and approaches developed by the OECD to assist policy makers and practitioners in this area. The report is focused on the financing of WSS rather than on the water sector as a whole.[1]

Structure of the report

The report is structured in two main parts: ***Part I*** provides a comprehensive overview of key issues with respect to financing the water and sanitation sector, while ***Part II*** presents a number of tools and approaches developed by the OECD to assist policy makers and practitioners in this area.

Part I is organised in three chapters, as follows:

Chapter 1 identifies the investments required to build, operate and maintain the infrastructure for providing sustainable water and sanitation services. It then examines the substantial benefits that WSS generate for human health, the economy as a whole and the environment. Access to clean drinking water and sanitation reduces health risks and frees up time for education and other productive activities, as well as increases the productivity of the labour force. Safe wastewater disposal helps to improve the quality of surface waters with benefits for the environment (*e.g.* functioning of ecosystems; biodiversity), as well as for economic sectors that depend on water as a resource (*e.g.* fishing, agriculture, tourism). Such benefits usually outstrip the costs of service provision and provide a strong basis for investing in the sector.

Chapter 2 assesses the current status of WSS and examines investment needs, identifies the financing sources and estimate the financing gaps to reach internationally agreed targets. In both OECD and non-OECD countries, the investment needs are huge and are unlikely to be met if current trends continue. This is particularly critical in the context of the current financial and economic crisis which is affecting financing sources for public and private investments alike.

Chapter 3 examines where the money is going to come from, including from a combination of efficiency gains, adjusted targets and additional financial resources. In the long run, increasing the "3Ts" – tariffs, taxes and transfers – would be the most sustainable way to close the financing gap. In the interim, however, repayable financing is likely to be needed to bridge the financing gap. Mobilising repayable financing calls for innovation, as the sector has traditionally not been able to attract much repayable financing, particularly when compared with other infrastructure sectors.

Part II describes and evaluates the tools developed by OECD to address the key financing issues described in Part I.

Chapter 4 introduces how the set of tools presented in the toolbox can help governments and water sector actors improve their policies and practices.

Chapters 5 to 10 contain a brief description of the tools on the basis of a common format. The tools presented in these chapters include the following:

- Strategic Financial Planning (at national or regional level): the FEASIBLE tool;
- Financial Planning Tool for Water Utilities;
- Multi-Year Investment Planning Tool for Municipalities;
- Guidelines for Performance-based contracts;
- Water Utility Performance Indicators (IBNET);
- Private Sector Participation – A Checklist for Public Action.

Note

1. A companion OECD report (OECD [2011], *Financing Water Resources Management*, Paris) examines issues relative to financing Integrated Water Resource Management.

Part I

Key issues with respect to financing water and sanitation

Chapter 1

What are the benefits of investing in WSS?

> *This chapter identifies the types of investments that are required to deliver WSS and presents available evidence on the magnitude of the benefits that are generated from such services. Such benefits usually outstrip the costs of service provision and provide a strong basis for investing in the sector.*

Water and sanitation services (WSS) generate substantial benefits for human health, the economy as a whole and the environment. Access to clean drinking water and sanitation reduces health risks and frees up time for education and other productive activities, as well as increases the productivity of the labour force. Safe wastewater disposal helps to improve the quality of surface waters with benefits for the environment (*e.g.* functioning of ecosystems; biodiversity), as well as for economic sectors that depend on water as a resource (*e.g.* fishing, agriculture, tourism).

For such benefits to be generated sustainably, investments in a whole range of services alongside the WSS value chain need to be carried out, ranging from protecting the raw material (freshwater resources) to building storage capacity or water transport networks, all the way to investments into collection, safe disposal, treatment or reuse of wastewater. Once built, the infrastructure needs to be adequately maintained and operated so as to provide sustainable, affordable and reliable access to water and sanitation services. New and recurrent investments in water and sanitation services are therefore critical in order to expand access to the services and maintain their ability to deliver benefits over time.

1.1. Identifying necessary investments in WSS

What investments are needed?

The report considers the investments needed to ensure sustainable provision of WSS services alongside the entire WSS "value chain", as shown in Figure 1.1.

Providing access is usually considered as the main entry point for the delivery of WSS. Access to water services can be provided via a well or a handpump or via a reticulated network system. When water is provided via a network, this can be done via a household connection (within the house or in the yard) or a public connection, referred to as standpipes or tap stands. Investments required can range from digging a well and maintaining it in good working order to building water transport and distribution networks with associated water treatment facilities.

To ensure that water is provided to the right standard (defined based on WHO guidelines on drinking water quality), water treatment is necessary to remove suspended solids, bacteria, algae, viruses, fungi, minerals and man-made chemical pollutants including fertilisers. Treatment is usually carried out off-site at the point of source, although it may also be required at point-of-use (*i.e.* at household level), as water may be contaminated during transport or storage. Examples of water treatment technologies include filtration, chlorination, flocculation, solar disinfection, boiling and pasteurising.

Figure 1.1. **The value chain of sustainable water and sanitation services**

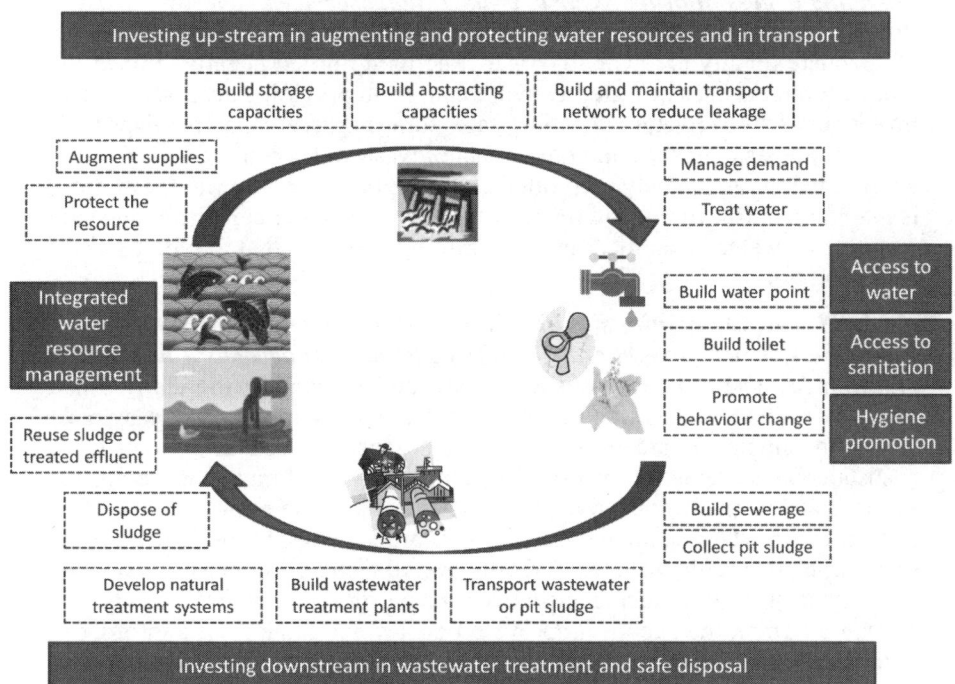

Source: OECD (2011), *Benefits of Investing in Water and Sanitation: an OECD Perspective*, OECD, Paris.

According to the WHO-UNICEF Joint Monitoring Program, "Sanitation is generally referred to as the provision of facilities and services for the safe disposal of human excreta. Sanitation also refers to the maintenance of hygienic conditions, through services such as garbage collection and wastewater disposal". Providing access to sanitation usually means investing in the first segment of the sanitation value chain, *i.e.* ensuring that humans are adequately separated from their excreta. There are two main types of facilities for collecting human excreta: on-site sanitation systems (such as dry-pit latrines or ventilated improved pit latrines) and network-based sanitation solutions, with or without treatment of the sewage collected.

Hygiene promotion is a key intervention to ensure that access to water and sanitation services can deliver benefits. They include provision of hand washing points, hygiene and health education and the encouragement of specific behaviours such as hand washing at critical times, keeping animals out of the kitchen, proper management of child excreta and proper storage of household drinking water.

Adequate investments are needed both downstream and upstream from providing access in order to ensure sustainable services. Investing in water resource management upstream is critical, so that sufficient water resources of adequate quality are available over time with limited negative impact on other alternative uses of water. Downstream from providing access, adequate investment in wastewater collection, safe storage or treatment and disposal is necessary so as to ensure that the impact of wastewater being released in the environment is adequately controlled and good quality of the water resources is maintained. Recycling and reuse of treated wastewater can also reduce the amounts of water consumed and generate by-products that can be used for agriculture or energy production.

WSS typically require significant capital investments up-front in long-lived assets, which can generate benefits over several decades if adequately maintained. The bulk of investments are underground (particularly piped networks), which means that monitoring asset condition is not an easy task. Relatively simple equipment, such as hand pumps, can also fall into disrepair if sustainable systems for ensuring ongoing repairs and maintenance are not in place. To maintain incentives for efficient service delivery, it is therefore critical to invest in adequate "sector software", alongside the hardware. At sector level, this could include improving overall sector governance, conducting tariff reforms or introducing incentives for performance improvements (see section 3.1 about the need to improve efficiency and reduce costs in order to shrink the sector's financing gap).

Who is responsible for investing?

Investors in water and sanitation services differ according to the type of services provided. For the most basic levels of service, such as a well, borehole or an on-site sanitation facility, households would be the prime investors. For anything beyond that, services are usually provided by a distinct "service provider". The organisation of water and sanitation service provision varies widely from one country to another and water service providers have different financing requirements and risk profiles. In about 90% of cases, formal water services are delivered by public entities, which may include state-owned enterprises, local governments, municipal companies, asset-holding agencies, etc. Ministries and government agencies are also primary investors in "sector software" and accompanying measures.

Water services are usually locally provided, given that water and sewage are bulky and costly to transport over long distances, with a limited case for integrated transportation networks as they exist for electricity or gas. As a result, most water service providers were initially set up at the municipal level. Over the years, however, market structure reforms in the water sector have oscillated between decentralisation reforms, which may be driven by broader country-wide decentralisation processes and some degree of aggregation (in order to reach a more efficient scale of operations). In some countries, particularly in the OECD, the pressure for achieving economies of scale for service provision has led to some degree of aggregation, either through the formation of groupings of municipalities (as in France, Italy or Spain) or the creation of regional or even national providers (such as watershed-based water companies in England and Wales, regional companies in Portugal and Italy, State-level companies in Brazil or national ones in West Africa). In developing countries, since the early 1990s, WSS have progressively been decentralised, which means that currently, the majority of water and sanitation service providers in those countries tend to operate at the local level. Such decentralised authorities have often been struggling to establish their financial standing in order to access financing on their own credit.[1]

As discussed in the rest of the report, all these entities can source financing from a variety of sources (including tariffs, transfers, ODA or repayable financing). However, differences in ownership or scale of operations can have a substantial impact on the type of financing that can be mobilised and at which cost.

1.2. Estimating the benefits of investing in WSS

Benefits from the provision of basic water supply and sanitation services are massive and far outstrip costs. In most OECD countries, these benefits have been reaped since the late 19th all the way through the late 20th century when basic water and sanitation infrastructure was extended to reach large parts of the population. For example, in In Marseille (France), water supply was a significant constraint on the city's growth during the early 19th century. A catastrophic drought in 1834 meant that water availability dropped from 75 litres per capita per day to 1 litre per capita per day and triggered a cholera epidemic. This in turn led to the construction of a canal to bring water, which allowed augmenting water supply to 370 litres a day after its completion in 1848. Increased water availability helped bring down mortality significantly, although it remained at much higher levels than in other French cities at the time (28 deaths/1 000 inhabitants as opposed to 9/1 000 in Paris at the same time). Higher water supply also meant more dirty water lying about: it is not until ambitious sewerage works were completed and households got connected to the sewers that mortality rates dropped significantly. Although attributing causality is always a perilous exercise, Figure 1.2 shows a clear correlation between a reduction in mortality and the timing of water and sanitation investments.

In France, overall, the total length of water supply networks grew from about 25 000 km in 1940 to over 800 000 km in 2004 (Smets, 2008). Only

Figure 1.2. **Impacts of water and sanitation investments on mortality in Marseille (France)**

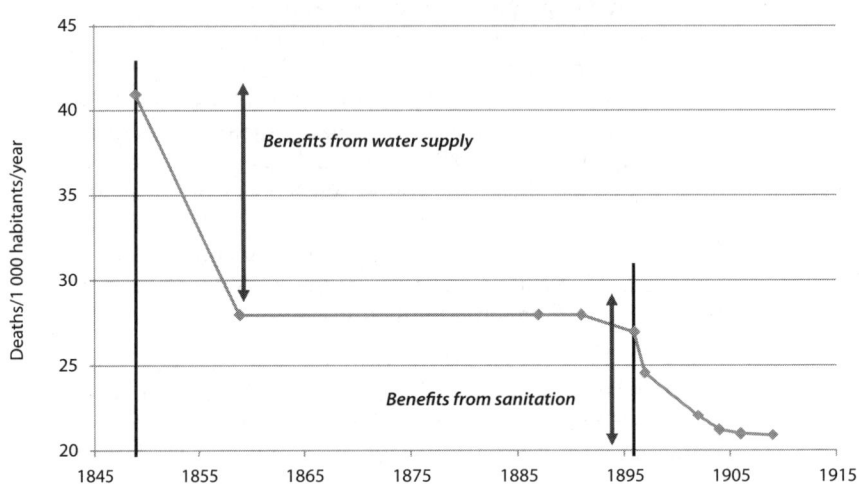

Source: AESN (2007), "Bénéfices de l'assainissement", Rapport d'étude AESN, February 2007, Seine Normandie, France.

27% of the French population had toilets inside their home in 1954 against 98% today and three quarters of the treatment plants in operation by 2009 was built after the 1990s (although the older ones tended to be larger plants). In the United States, the introduction of water chlorination and filtration in 13 major US cities during the early 20th century led to significant reductions in mortality with a calculated social rate of return of 23 to 1 and a cost per person-year saved by clean water of about USD 500 in 2003 US dollars.[2]

In developing countries, WHO has estimated that almost 10% of the global burden of disease could be prevented through water, sanitation and hygiene interventions. Children are most affected, with 20% of disability adjusted life-years (DALYs)[3] in children under 14 attributable to inadequate water, sanitation and hygiene and 30% of deaths of children under 5.

Health benefits from improved access to sanitation and hygiene appear to be most significant, followed by improved access to clean water. With respect to water, there is reasonable evidence to support the finding that the quantity of water provided is paramount (particularly in order to adopt basic hygienic practices) if health benefits are to be achieved and may be more critical than the quality of such water, which is also important.

In developing countries, WHO estimated that achieving the MDGs for water and sanitation could generate an estimated USD 84 billion per year in

Table 1.1. **Overall benefits of meeting the MDGs in water and sanitation**

Type of benefits	Breakdown	Monetised benefits (in USD)
Time savings from improved water and sanitation services	• 20 billion working days a year	USD 63 billion a year
Productivity savings	• 320 million productive days gained in the 15-59 age group	USD 9.9 billion a year
	• 272 million school attendance days a year	
	• 1.5 billion healthy days for children under 5	
Health-care savings		USD 7 billion a year for health agencies
		USD 340 million for individuals
Value of deaths averted, based on discounted future earnings		USD 3.6 billion a year
Total benefits		**USD 84 billion a year**

Source: Prüss-Üstün et al., 2008, *Safer water, better health: costs, benefits and sustainability of interventions to protect and promote health*, World Health Organization, Geneva, 2008, based on an evaluation by Hutton and Haller (2004).

benefits, with a benefit to cost ratio of 7 to 1.[4] As shown in Table 1.1, three quarters of the benefits would stem from time gains, *i.e.* time that is gained by not having to walk long distances to fetch water or to queue at the source.[5] Most other benefits are linked to a reduction of water-borne diseases such as reduced incidence of diarrhoea, malaria or dengue fever, which are estimated either in terms of reduced health care costs or productivity savings.

In addition, WSS generate a number of non-economic benefits that are difficult to quantify but that are of high value to the concerned individuals in terms of dignity, social status, cleanliness and overall well-being. More broadly, adequate water and sanitation services appear to be a key driver for economic growth (including investments by firms that are reliant on sustainable water and sanitation services for their production processes and their workers).

Wastewater collection and treatment can generate health and environmental benefits, with ripple effects on other economic sectors, such as agriculture, fisheries, tourism or industry. The benefits of wastewater collection and the resulting protection from contamination are obvious to most individuals. By contrast, the benefits of wastewater treatment are less obvious to individuals (as is often the case with public goods) and more difficult to assess in monetary terms. The consensus on the need for increased urban wastewater treatment as well as safe disposal of its residues has therefore developed more slowly, probably also due to the relatively high costs of such interventions. In the United States, the 1972 Clean Water Act built an important legal basis for expanding wastewater treatment facilities. In Europe, the European Union Urban Waste Water Treatment Directive adopted in 1991 represented the policy response to the growing problem of untreated sewage disposed into the aquatic environment.

All benefits from wastewater treatment are linked to an improvement in water quality through the removal of different polluting substances, generating withdrawal benefits (*e.g.* for municipal water supply as well as irrigated agriculture, livestock watering and industrial processes) and in-stream benefits (benefits that arise from the water left "in the stream" such as swimming, boating, fishing).

Wastewater treatment can have a beneficial impact on the environment and economic activities that are dependent on it. For example, in the Black Sea, the degradation of water quality due to enrichment in nutrients led to an important increase in algal mass affecting aquatic life. The mass of dead fish was estimated at around 5 million tons between 1973 and 1990, corresponding to a loss of approximately USD 2 billion. Water quality is also an essential factor for certain tourism activities and sewage treatment leads to enhanced tourism attraction. In most countries, non-compliance with certain norms for bathing water leads to the closure of beaches and lakes for recreational purposes and therefore influences strongly the local tourism economy. In Normandy (France), for example, it has been estimated that closing 40% of the coastal beaches would lead to a sudden drop of 14% of all visits, corresponding to a loss of 350 million Euros per year and the potential loss of 2 000 local jobs.

Benefits for property have also been shown to be significant. People living in the surroundings of water bodies benefit from increased stream-side property values when wastewater treatment measures ensure a certain quality of water bodies. Several studies show that in proximity of areas that benefited from improved water quality, property values were found to be 11% to 18% higher than properties next to water bodies with low quality.

Finally, wastewater treated to adequate levels can be reused. Both faeces and urine can be used as potent fertilisers for agriculture, as well as for producing biogas for energy production. For example, biogas plants can be built to use animal and human waste to produce a colourless clean gas similar to liquefied petroleum gas (LPG), which can be used for cooking and lighting with virtually smoke-free combustion. A study by Winrock International evaluated an integrated household-level biogas, latrine and hygiene education programme in Sub-Saharan Africa and found that the programme's economic rate of return was 178%, with a 7.5% financial rate of return.[6] Sludge from wastewater treatment plants can also be mixed with biodegradable municipal waste. However, making such projects economically viable would require operating at a large scale, with waste collected from several hundred thousand people.

Aggregated economy-wide assessments of benefits of water quality improvements are very few and far between, however. The US Environmental Protection Agency estimated the net benefits of water pollution legislation in the last 30 years in the United States at about USD 11 billion annually, or about USD 109 per household. In South East Asia, the Water and Sanitation Program estimated that, due to poor sanitation, Cambodia, Indonesia, the Philippines and Vietnam lose an aggregated USD 2 billion a year in direct financial costs (equivalent to 0.44% of their GDP) and USD 9 billion a year in economic losses (equivalent to 2% of their combined GDP). The financial losses include change in household and government spending as well as impacts likely to result in real income losses for households (*e.g.* health-related time loss with impact on household income) or enterprises (*e.g.* fisheries). The economic costs include the financial costs as well as longer-term financial impacts (*e.g.* less and fewer educated children, loss of working people due to premature death, loss of usable land, tourism losses) (Hutton *et.al.*, 2008).

Protecting the quality of the resource and balancing supply and demand so as to ensure water security and reliability are critical to ensure that benefits from WSS are generated sustainably over time.[7] For water services to be provided sustainably over time, it is critical to ensure that the raw material, clean water, is adequately protected and managed. This will become increasingly relevant with increasing pressures on the resource exerted by economic and demographic growth as well as the potential impacts of climate change on the water cycle.

Protecting water catchments and reducing pollution to water resources result in similar benefits to end-customers as those described from providing access to safe water. In addition, protecting water resources directly at the source by limiting pollution from catchments generates indirect benefits, such as avoided (investment and treatment) costs and can be overall more cost-effective, as discussed in Box 1.1. Increasingly, countries are recognising the benefits of managing water resources using a river basin approach, given that reducing pollution at the source tends to be a cheaper option than treating water before supplying it to consumers.

In order to ensure a reliable water supply, there is a need to balance water supply and demand. The degree of certainty with which water is supplied is an important factor in determining the benefit that water users derive from the service and strongly influences their willingness to pay. Increased reliability of water supplies avoids the need for households to store water for shortage situations and therefore induces cost savings. Water supply reliability is also an important parameter for economic activities (industries, but also agriculture and services) which use water in their processes or as a non-substitutable input.

Box 1.1. **Water catchment protection in New York (United States)**

The most famous case of benefits linked to water catchment protection is reported in New York. A new drinking water regulation required water suppliers to filter their surface water supplies, unless they could demonstrate that they had taken other steps – including watershed protection measures – to avoid harmful water pollution. Confronted with the choice between the provision of clean water through a newly built filtration plant or managing water sheds, New York City concluded that the latter was more cost-effective. Whereas the costs of the filtration plant have been estimated at between USD 6 billion to USD 8 billion, watershed protection efforts, including the acquisition of critical watershed lands and a variety of other actions designed to reduce contamination sources in the watershed, were estimated to cost only around USD 1.5 billion – thus four to five times less. As a consequence, New York City chose the second solution that favored investing in natural rather than built capital.

Source: Salzman (2005), as presented in OECD (2010b).

Figure 1.3. **The water and sanitation benefits curve**

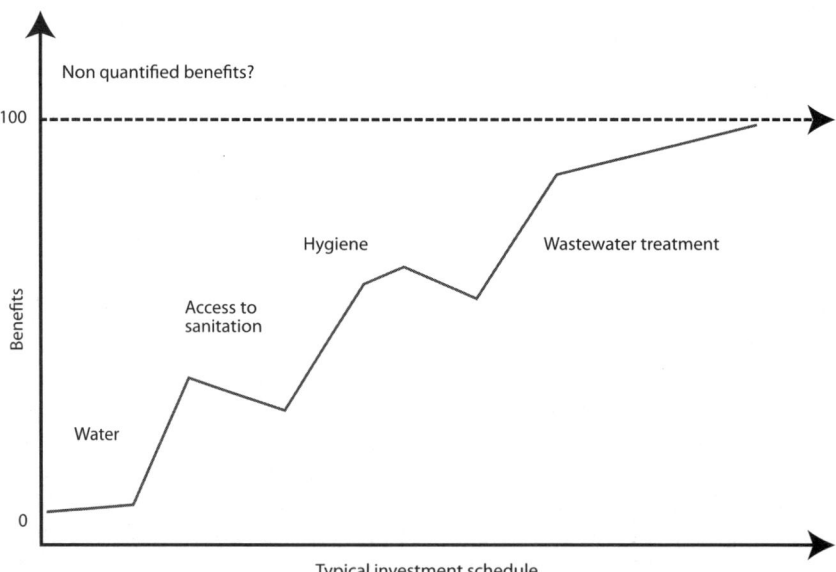

Source: OECD (2011), *Benefits of Investing in Water and Sanitation: an OECD Perspective*, OECD, Paris.

The combined magnitude of the benefits of WSS can vary substantially depending on the level of sector development. Figure 1.3 represents the streams of benefits coming from a typical investment schedule.

In most countries where the "access gap" is still large, providing access to water services is seen as a priority as it can indeed deliver substantial benefits, particularly if combined with hygiene education. If access to water is provided without corresponding investments in sanitation, however, this can generate temporary disbenefits, as abundant water supply can create pools of stagnant waters mixing with excreta and other types of waste (such as grey waters).

Connecting people to sewers without wastewater treatment can sometimes generate disbenefits in the cases in which it transforms diffuse pollution into point-source pollution (a sea outfall for example). Wastewater treatment would eliminate all residual risks. However, benefits would start tailing off once a high degree of wastewater treatment is reached (although this would clearly depend on maintaining existing installations, so that they can continue to deliver benefits). Going further, there may be some additional benefits (such as from an improved living environment or benefits for future generations) which may be harder to quantify but that could nevertheless justify investments in WSS beyond the level at which quantifiable benefits overtake costs.

There are few aggregated estimates of benefits and few rules of thumb that could be applied universally, however, given that benefits from water and sanitation investments tend to vary substantially according to local factors, such as the level of development of the infrastructure, the prevalence of water-related diseases, environmental status, etc. The benefits and costs of each particular investment or set of measures would therefore need to be estimated in each location specifically, in order to select the investments with the highest benefit-cost ratio and allocate scarce resources to the most cost-effective investments. Given that carrying out a full evaluation of benefits is potentially expensive and time consuming, one alternative from a methodological point of view is to compare interventions based on cost-effectiveness criteria, *i.e.* to evaluate how much different interventions cost in order to achieve similar objectives (and therefore generate the same amount of benefits).

In developing countries, for example, it was found that investing in WASH (water, sanitation and hygiene) is very cost-effective. The Disease Control Priority project (an ongoing effort to assess disease control priorities and produce evidence-based analysis and resource materials to inform health policymaking in developing countries) found that hygiene and sanitation promotion activities cost respectively USD 3 and USD 11 per DALY averted. By comparison, the cost-effectiveness of promoting oral rehydration therapy, the main other measure to prevent diarrhoea mortality, was estimated at USD 23 per DALY, which means that hygiene and sanitation promotion compares favourably to such measure. Infrastructure investments had a much higher cost when compared to effectiveness. For example, the cost-effectiveness of constructing sanitation facilities (including promotion) was USD 270 per DALY. As for water supply, providing a community connection was estimated to cost USD 94 per DALY, while it was more than twice as much for household connections (USD 223 per DALY). These measures are still cost-effective when compared to other health measures: for example, the provision of antiretroviral therapy against AIDS was estimated to cost USD 922 per DALY.[8]

Haller *et al.* (2007) conducted a cost-effectiveness analysis which indicated that the provision of in-house piped water supply and sewer connection is the intervention that maximises health gains but is also the most expensive intervention: they estimated that piped water supply and sewer connection would achieve a maximum health gains (71 million DALYs averted) but that investment and recurrent costs would also be quite important (ranging from USD 48 billion to 60 billion). From this analysis, they concluded that for many developing countries, in-house piped water supply may not be affordable in the short to medium-term and governments and households may need to settle in the short-term for second-best solutions, although health and non-health benefits would not be as large. They suggested that disinfection at point of use, which has a better cost-benefit ratio (USD 338 to USD 461 million for 17-19 million DALYs averted) could be used as an efficient short-term policy strategy to further reduce diarrhoea incidence, while time elapses during the extension of coverage and upgrading of piped water and sewage services. This investment strategy for water improvements is also recommended by Edwards (2008), in a guide to understanding costs and benefits of water interventions published by WHO.

In OECD countries, maintaining existing infrastructure so that it can continue to deliver an ongoing benefit stream should be a priority, together with the need to adapt to a changing climate, therefore generating substantial investment needs going forward. By contrast, the case for investing in additional improvements is sometimes less clear-cut, as the marginal value of incremental improvements is likely to go down, at least when measurable benefits are accounted for.

Notes

1. See OECD (2010a) for more detail on the issues faced by decentralised entities to access financing.

2. See OECD (2011) for more detailed facts and figures.

3. The sum of years of potential life lost due to premature mortality and the years of productive life lost due to disability.

4. MDG 7c calls for reducing by half the proportion of people without sustainable access to safe drinking water and basic sanitation (see *www.undp.org/mdg/goal7.shtml*).

5. The value of those time gains is estimated on the assumption that any time "gained" in such a way could be productively used on income-generating activities. This may not be possible in some developing economies, but such time would nevertheless be gained for other activities generating more intangible benefits, such as furthering education for children and adults alike.

6. Renwick, M. *et al.* (2007).

7. The benefits (and associated financing) of investing in integrated water resource management are discussed in more detail in a companion report (OECD [2012 forthcoming], *Financing Water Resources Management*, Paris).

8. Cairncross and Valdmanis (2006): *www.dcp2.org/main/Home.html*.

Chapter 2

Current status of WSS and investment needs

> *This chapter synthesises aggregated investment estimates in order to maintain and expand drinking water supply and sanitation services (WSS) around the world. In doing so, the Chapter distinguishes between the situation in countries with almost universal water and sanitation coverage (most OECD countries and some transition countries) and those where extending access to the service remains at the heart of current investment policies.*

Substantial investments are needed in order to deliver expected benefits from WSS. Key challenges include increasing access to water and wastewater services (mainly in developing countries but also in some OECD countries), replacing and maintaining ageing infrastructure and addressing water security and environmental concerns. Throughout the world, the challenges of providing access to safe water and sanitation are further accentuated by increasing demands from other water uses due to factors such as population increase, agricultural water needs for food production, rapid urbanisation, degradation of water quality, and increasing uncertainty about water availability, in part due to climate change.

Addressing these challenges will require large capital investments for new infrastructure as well as financial allocations for maintaining, repairing, upgrading and operating existing facilities. Indeed, most scenarios of future expenditure tend to focus on "investment" needs and exclude recurring expenditures for operations, maintenance, repairs, replacement and overhead. While these expenses are sometimes covered by revenue, cash shortfalls resulting from tariffs being set below costs often lead to inadequate expenditures for operations and maintenance, and a resulting increase in future investment needs.

2.1. Current status and investment needs in OECD countries and transition economies

In OECD countries, access to safe water supply and sanitation has largely been ensured following substantial investment over many decades (OECD, 2009a). As a result, in most OECD countries, 100% of the population has access to safe drinking water. With few exceptions, water supplied to the main population centres is bacteriologically safe (OECD, 2006a). In some countries, however, such as Mexico, New Zealand, Poland, Turkey or some parts of the United States, a segment of the population is not yet connected to networked water supplies, especially in rural areas.

With regard to sanitation services, considerable variations exist among OECD countries in terms of coverage and level of treatment. Some countries are still completing sewerage systems or implementing the first generation of municipal wastewater treatment plants, including Belgium, Mexico, and Turkey, which have however all made significant progress in recent years. Japan, Korea, Luxembourg, and the United Kingdom exhibit high secondary treatment coverage. The countries that have a particularly high level of tertiary treatment include Spain, Austria, Denmark, Finland, Germany, Netherlands, Sweden and Switzerland (OECD, 2009a). Regarding villages under 2 000 inhabitants, Spain, is working on non conventional wastewater treatments, which could also be useful for developing countries.

Despite a high initial asset base, developed countries confront huge costs of modernising and upgrading their systems, so as to comply with increasingly stringent health and environmental regulations, maintain service quality over time, ensure the security of water supplies in response to climate change, pollution and growing populations, and in some cases, overcome the neglect and underfinancing of earlier years. For example, according to a recent report by Conviri, the agency monitoring water resources in Italy, the country will need to invest some EUR 50 billion in the water and wastewater sector over the next 20 years, with particular needs in terms of reducing leakage and investing in wastewater treatment.[1]

According to OECD (2006a), the global capital costs of maintaining and developing WSS infrastructure in OECD countries plus the BRICs could amount to 0.35 to 1.2% of their GDP. This corresponds to total projected annual needs of around USD 780 billion by 2015 and USD 1 037 billion by 2025, up from a current estimated expenditure on water infrastructure of USD 576 billion annually. According to OECD (2007), this is far higher than comparable estimates for roads (USD 160 billion per year by 2020) or electricity transmission and distribution (around USD 80 billion per year by 2025).

This report highlighted that there is a wide range of estimates of required annual expenditures in the water sector, however, depending on the methods used for evaluation. The report stressed the wide variations from region to region reflecting very different levels of infrastructure coverage and economic ability (or political will) to take account of environmental pressures. The headline figures were estimated based on the review of investment needs in a number of OECD and non-OECD countries, which concluded that going forward, the levels of expenditure on water services for high income countries should be of the order of 0.75% of GDP (ranging between 0.35% and 1.2%) and could go up to 6% for some low-income countries which need to cover previous investment deficits in the sector. As illustrations, France and the United Kingdom will have to increase their water spending as a share of gross domestic product (GDP) by about 20% to maintain water services at current levels; Japan and Korea may have to increase their water spending by more than 40%. Finally, it noted that most estimates tend to focus on investments and ignore the need to cover the costs of operations and maintenance.

Lloyd Owen (2009) sought to derive more comprehensive estimates by forecasting spending needs both for investments and operations and maintenance, across a large number of countries, across developed and developing ones. They estimated that meeting future challenges (such as rehabilitating existing assets or meeting the MDGs) would call for around USD 2 880 billion in investments over the next two decades (or about USD 144 billion per year) in the 67 countries covered by the analysis, with associated operating costs which could be twice as high as capital investment costs, as shown in Table 2.1. This report also identified a substantial financing gap as it estimated that only USD 631 to 1 381 billion could be generated from existing sources of revenues (including tariffs), leaving a gap of between USD 1 049 billion to 2 297 billion over the period.

Table 2.1. **Forecast operating and capital spending in countries covered, 2010-29 (USD billion)**

	Operating costs	Capital spending (capex)			% capex by region
		Low	Medium	High	
North America	1 821	525	630	940	23%
Europe	2 133	642	838	991	28%
Developed Asia	1 018	461	550	640	19%
Latin America	796	119	164	194	5%
Rest of World	992	472	713	1 027	24%
Overall	**6 760**	**2 213**	**2 880**	**3 792**	**100%**

Source: Lloyd-Owen, D. (2009), *Tapping liquidity: financing water and wastewater to 2029, a report for PFI market intelligence*, Thomson Reuters, London.

In EECCA countries (Eastern Europe, Caucasus and Central Asia), the need for maintaining and upgrading existing infrastructure is combined with sometimes significant needs to expand coverage and address the challenges of poor governance, institutional inefficiency and the deterioration of the asset base. Much of the existing infrastructure is old and over-sized for present needs, and is ill-suited to economic and demographic realities. A number of these countries cannot afford to maintain even existing services in their present form, and face a situation where they have to choose between maintaining affordable tariffs and skimping on quality by lowering the standards of service. OECD (2009a) refers to the examples of Armenia, Moldova or Georgia where the current levels of financing are clearly insufficient even to maintain assets at their present low operational levels or to provide adequate levels of service, with the corresponding long-term cost impacts. In the Commonwealth of Independent States, JMP (2010) found that the rate of access to piped water in the home has declined by 2% between 1990 and 2008 (from 71% to 69%), which points to clear under-investment in the sector. In addition, OECD (2006b) pointed out that JMP figures paint an over-optimistic picture of the situation with respect to access to water and sanitation services in the region. In many EECCA countries, a sharp deterioration in service levels implies that "having a water tap does not necessarily mean having sustainable access to safe drinking water". Cross-contamination between water and sewerage networks, due to high levels of leakage, for example, can have serious effects on public health. To meet the MDGs in EECCA countries, it was estimated in 2006 that EUR 7 billion would be necessary annually for operation, maintenance and capital investments, which was roughly double available financing at the time.

In Moldova, for example, various alternative policy targets have been costed, ranging from a baseline scenario (with a halt to the deterioration of existing infrastructure, modest improvements, and more spending on O&M) to complete fulfilment of the draft Government strategy (including compliance with EU Directives, achievement of the MDGs, construction of certain critical wastewater treatment plants, etc). Depending on the sector target chosen, the total 20-year costs ranged from EUR 1.3 billion to 3.2 billion for a total population of just over 4 million people (EUR 325-EUR 800 per capita).[2]

2.2. Overview of investment needs in developing countries: Reaching the MDGs

In developing countries, a significant percentage of the population still does not have access to water and sanitation services, while many others suffer from unsatisfactory services. The international community is committed to achieving the Millennium Development Goals (MDGs) that aim to halve the proportion of people without access to safe drinking water and basic sanitation by 2015. Despite strong calls for action at international level, the Joint Monitoring Program, led by WHO and UNICEF, found that 2.6 billion people still do not use improved sanitation (out of which 1.1 billion still defecate in the open), while 884 million people do not use improved sources of drinking water (JMP, 2010).

The situation looks fairly positive on the drinking water supply front. At present, 87% of the global population uses improved drinking water sources, which represents a 10% increase since 1990. At the current rate of progress, the world is expected to exceed the MDG target, although it will mean that 672 million will still not have access to improved water sources by 2015. This substantial improvement has largely been driven by good performance in the two most populous countries, India and China: nearly half the people who gained access to improved water over that period live in those two countries.

By contrast, a number of countries are unlikely to meet the MDG for drinking water supply, particularly in Sub-Saharan Africa where 37% of the population without access to drinking water is located and progress has been particularly slow. In addition, it must not be forgotten that individuals need and expect a better access to water than what is currently defined as "access to improved water sources". In particular, according to the human right to safe drinking water and sanitation adopted by the UN General Assembly in July 2010 (Resolution 64/292), individuals need access to water that is safe, accessible, acceptable and affordable. None of these characteristics is measured by the current MDG indicator.

The global figures hide substantial regional differences and within individual countries. Urban-rural disparities are particularly striking, as 84% of the people without access to improved drinking water live in rural areas (JMP, 2010). In Russia, for example, Martoussevitch (2008) reported striking differences in the levels of capital investment from one region to another. In 2006, the author calculated that investments in water supply services in the regions that were best performing economically were 15 390 times higher than in the worst performing. This disparity in investment was much higher than the disparity in per capita GDP. Furthermore, the number of people who have gained access to improved water sources in urban areas between 1990 and 2008 has been lower than population growth, resulting in a net decrease in urban access rates. By contrast, gains in terms of rural water coverage have exceeded population growth in the same areas.[3]

Much more progress still needs to be made on the sanitation front. The world as a whole is off-track with regard to the sanitation target: at the current rate of progress, the world will miss the MDG target by 13 percentage points, which means that by 2015, it is projected that 2.7 billion will not have access to improved sanitation and that 1 billion people, who should have benefited from MDG progress, will miss out (UN-Water, 2010). Almost three quarters of those who live without access to improved sanitation live in Southern Asia, but there are also large numbers in Sub-Saharan Africa (JMP, 2010). Seven out of 10 people without access to improved sanitation live in rural areas.

There are a number of issues with the MDG indicators as they are currently defined and measured. For example, access to water-supply services is defined as having access via an "improved" source. In Sub-Saharan Africa, however, one third of the trips to improved water sources take more than 30 minutes, which means that people collect considerably less water than would be necessary to adopt safe hygienic practices. According to Bartram (2008), providing water in the home would be much preferable in order to protect health and secure social benefits. Raising the indicator to such a standard would mean missing the target on the water front as well, however. Another issue raised by the JMP itself is the difficulty and high cost of measuring whether or not the water is safe to drink.

As the target date for the MDGs is drawing nearer, a debate has been initiated on the indicators that may be appropriate to use for the sector beyond 2015. This needs to take into account the recent adoption of the human right to safe and clean drinking water and sanitation in July 2010, as stated in Box 2.1.

There is a broad range of estimates for the costs to reach the MDGs, depending on the assumptions used on the types of investment made. According to the GLAAS report, the global cost estimates for meeting the

> Box 2.1. **The human right to safe and clean drinking water and sanitation**
>
> With respect to the human right to safe and clean drinking water, the report states that the following characteristics should apply:
>
> (a) **Sufficient quantity.** Water must be available in a quantity sufficient to satisfy all personal and domestic needs;
>
> (b) **Water quality.** It must not pose a threat to human health. The World Health Organization Guidelines for Drinking-water Quality serve as an important reference in this regard;
>
> (c) **Regularity of supply.** Water supply must be sufficiently reliable to allow for the collection of amounts sufficient to realise all personal and domestic needs over the day;
>
> (d) **Safety of sanitation facilities.** Human, animal and insect contact with human excreta must be effectively prevented. Regular maintenance, cleaning and – depending on the technology – emptying is necessary to that extent. Sludge and sewerage must be properly disposed of to avoid negative impacts on water quality and human health;
>
> (e) **Acceptability.** Sanitation facilities, in particular, must be culturally acceptable. This will, for instance, often require privacy as well as separate male and female facilities when these are shared;
>
> (f) **Accessibility of services.** Services must be available within or in the immediate vicinity of each household as well as schools, workplaces, health-care settings and public places. Access must be ensured in a sustainable manner;
>
> (g) **Affordability of services.** Regulation also has to set standards regarding pricing. Water and sanitation services do not have to be provided for free and tariffs are necessary to ensure the sustainability of service provision. To meet human rights standards, the essential criterion is that tariffs and connection costs are designed in a way, including through social policies, that makes them affordable to all people, including those living in extreme poverty.
>
> *Source:* Human Rights Council A/HRC/15/31, Report of the independent expert on the issue of human rights obligations related to access to safe drinking water and sanitation, Catarina de Albuquerque, 29 June 2010, pp. 16-17, para. 47.

drinking water and sanitation MDG target range from USD 6.7 billion to USD 75 billion per year, *i.e.* USD 33.5 billion to USD 375 billion by 2015 (UN-Water, 2010). There is a ten-fold variation in the cost estimates, largely due to the fact that estimates are based on different assumptions with respect to baseline years, population growth, cost of technology and levels of service.

Some of the cost estimates include only the cost of new capital infrastructure and do not consider the costs of maintaining or rehabilitating existing infrastructure, which can be very significant. For example, Hutton and Bartram (2008) estimated spending required to meet the MDG target at USD 42 billion for water and USD 142 billion for sanitation, a combined annual equivalent of USD 18 billion. The cost of maintaining existing services totals an additional USD 322 billion for water supply and USD 216 billion for sanitation, a combined annual equivalent of USD 54 billion. In addition, administrative costs, incurred outside the point of delivery of interventions, of between 10% and 30% were estimated necessary for effective implementation. A report by Hutton and Bartram (2008) highlights that 75% of annual needs to attain the MDG target for water and sanitation relate to the maintenance and the replacement of existing infrastructure, while 20% relates to the extension of sanitation services and 6% of water services. While the need for capital investment for new systems is often emphasised, there are significant costs associated with human resources and operation and maintenance to ensure that the existing systems are kept functional. According to Fonseca and Cardone (2005), most estimates do not appear to include the costs of support services or institutional capacity to ensure that systems are planned, installed and maintained adequately.

Current financing allocations will not be sufficient to meet the MDGs or even more, to guarantee the human right to safe and clean drinking water and sanitation. Despite the clear benefits for human and economic development (see Chapter 1), insufficient resources are currently allocated to meet the Millennium Development Goal (MDG) targets for sanitation and drinking water (in some countries). OECD (2009a) highlighted that what appears to be needed is roughly a doubling of the annual rate of investment.

WHO conducted a survey (in the context of the GLAAS report, UN-Water 2010) and asked whether Governments deemed that current financial flows were sufficient to meet the MDG target: 35 out of 37 reported that insufficient funding had been allocated for the sanitation target. Despite a substantial investment backlog in sanitation, the sub-sector receives a comparatively lower share of spending. The GLAAS report estimated that spending on sanitation accounted for about 37% of development aid to the sector as a whole. Among the countries that were able to separate out water from sanitation spending, sanitation accounted for about 20% of total spending on water and sanitation. The financing gap for meeting the MDGs is particularly acute in Sub-Saharan Africa, as recently analysed in the context of the Africa Infrastructure Country Diagnostic (see Box 3.1 in the next chapter for more information on this evaluation and potential ways to close the gap).

Notes

1. Global Water Intelligence, "Italy's €50 billion investment gap", Vol.11, Issue 8 (August 2010).
2. OECD/EAP Task Force (2008).
3. JMP (2010).

Chapter 3

Where is the money going to come from?

> *This chapter presents all possible sources of finance in turn and evaluates the potential for generating additional financing from each of these sources. In doing so, we examine the likely impact of the ongoing financial and economic crisis. The chapter also examines the potential role of the private sector in helping mobilising financing for the sector.*

Closing the financing gap will require countries to mobilise financing from a variety of sources, which may include reducing costs (via efficiency gains or the choice of cheaper service options), increasing the basic sources of finance that can fill the financing gap, *i.e.* tariffs, taxes and transfers (commonly referred to as the "3Ts") and mobilising repayable finance, including from the market or from public sources, in order to bridge the financing gap. These potential sources are shown in Figure 3.1.

Figure 3.1. **Sources of finance for WSS**

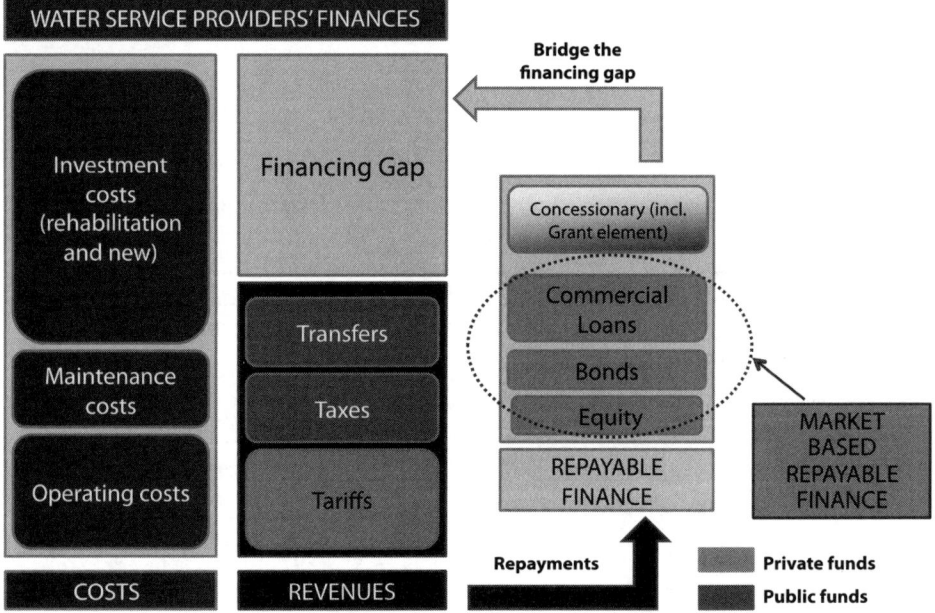

Source: OECD (2010), *Innovative Finance Mechanisms for the Water Sector*, OECD, Paris.

According to OECD (2009b), defining how these various sources of finance can be combined should be done based on Sustainable Cost Recovery (SCR) principles. SCR entails securing future cash flows from a combination of the 3Ts, and using this revenue stream as the basis for attracting repayable sources of finance – loans, bonds and equity, depending on the local situation. This is a key departure from earlier concepts of Full Cost Recovery (FCR) which implied that tariffs alone should be sufficient to cover all costs. In practice, particularly in poor countries where affordability is a significant constraint, SCR implies that public spending will often be required to complement revenues from tariffs, at least for a transition period.

Closing the financing gap will require countries to make efforts both on the "demand side" and the "supply side" of the sustainable finance equation (OECD, 2009a). On the demand side, the costs of providing WSS services can be reduced through improved investment planning and operating efficiencies. On the supply side, additional revenue sources can be mobilised from the 3Ts or from repayable sources through making the case for investment in WSS, improving the allocation of resources or reducing risks to attract private investments. Planning for the right balance between all these sources of revenues calls for strategic financial planning, so as to evaluate the potential for mobilising financing from each source of revenues as well as reducing costs.

Information on some of these financing sources tends to be patchy, however, which makes it difficult to reliably evaluate the gap between needs and available funding. For example, some financial information is available for central government and external donors spending, but information on subnational and local government expenditures is seldom aggregated at a national level. In addition, because funding for sanitation and hygiene is often spread over several different institutions, budget data are less available for sanitation and hygiene than for drinking water. Data on private sector investments (ranging from large private operators, informal providers, households or remittances) is notoriously difficult to collect, although they potentially represent an important source of funding for the sector.

3.1. Reducing costs and improving efficiency

Reducing the costs and improving the efficiency of existing water systems can be a crucial way of generating financial resources (as well as preserving physical resources, particularly in areas of water scarcity). Furthermore, selecting cheaper and more locally appropriate investment options or adapting service levels can generate substantial savings and support the definition of more realistic investment programs.

Reducing costs: Improving the efficiency of operations

Inefficiencies are responsible for important losses of funds within the sector. Operational inefficiencies include poor revenue collection, distribution losses (referred to as leakage or non-revenue water, NRW), labour inefficiencies and petty corruption.[1] For example, reducing NRW can significantly reduce operating costs, because it generates savings in terms of lower amounts of water used, reduced treatment and transport costs (as moving the water around can use a substantial amount of energy). Accumulated inefficiencies and deferred maintenance can result in higher costs over time. The Africa Infrastructure Country Diagnostic (Banerjee, 2011) estimated that, in Sub-Saharan Africa, inefficiencies of various kinds generated a cost to the sector of an average 0.5% of GDP (or USD 2.9 billion a year), and could rise up to 1.2% of GDP for low income fragile states (although they include in such inefficiencies the fact that tariffs are charged below cost-recovery levels).

They identified three main ways to reduce such inefficiencies, by raising user charges closer to cost-recovery levels (to provide more efficient price signals and help capture lost revenue), reducing utilities' operating inefficiencies (to prevent waste of significant resources, support healthier utilities, and improve service quality) and by improving budget-execution rates. They estimated that if such inefficiencies were eliminated, the funding gap to meet the MDGs could be almost eliminated in middle-income countries, even though it would still remain substantial in other countries (the majority) in Sub-Saharan Africa.

In Greater Cairo (Egypt), a Strategic Financial Planning exercise conducted with the support of the European Union Water Initiative (Mediterranean Component) and the OECD found that a series of efficiency measures, including reducing domestic consumption, reducing water losses and improving pumping efficiency would allow lowering overall system costs by 19% but that this would only make a minor contribution to reducing the financing gap faced by the city to maintain existing assets and meet future needs. If no measures are taken, the financing gap is expected to increase by 45% between 2006 and 2026 due to very low user charges, a serious backlog of investment accumulated over the past decades and a strongly projected demographic growth over the next 20 years.

The scope for realising efficiency gains is particularly high in developing countries and EECCA countries. As mentioned in OECD (2009a), whereas leakage rates are typically in the range of 10 to 20% in OECD countries, they frequently exceed 40% and sometimes reach up to 70% in developing country utilities. In Armenia, for example, OECD/EAP Task Force (2007) identified that water losses could go up to 70% in certain cities due to extensive leakage in the worn-out public networks and buildings' internal piping, excessive pressure in the water supply network or defective meters. The high rate of leakage in many systems is one, highly visible, aspect of the more general problem of inefficient operations.

In developing countries, electricity and chemical consumption tends to be high as well as the ratio of staff per connections. State-owned enterprises are often used as social buffers to (very inefficiently) transfer rents or resources to the population. African utilities have an average of five employees per 1 000 connections, more than twice the two employees per 1 000 connections frequently used as the international benchmark for developing countries (Ghosh Banerjee and Morella, 2010). Poor commercial performance (*i.e.* delays in collecting bills or the accumulation of bad debts) can also lead to cashflow problems, even if tariffs are set at a level that should be sufficient to cover costs.

Some OECD countries are also faced with comparatively high levels of leakage, particularly when the networks were installed several decades ago and have reached the end of their useful life. This is the case in London, for example, where the first comprehensive water networks were laid in the Victorian era. High leakage rates (above 40%) prompted the economic regulator, Ofwat, to introduce compulsory leakage reduction targets, which in turn triggered investment in a comprehensive pipe replacement programme. As discussed in OECD (2009b), however, there are decreasing returns from continuous leakage reduction efforts, given the existence of an "economic level of leakage", *i.e.* a rate at which it would cost more to make further reductions than to produce the water from another source.

There are many potential ways to stimulate increases in efficiency. Incentives for improved efficiency can be introduced with a number of tools, including price regulation, assignment of risks and rewards, competitive tendering, penalties and benchmarking. As a first step, benchmarking tools, such as IBNet (see chapter 9), can be used to compare the performance between various utilities and identify areas of potential inefficiencies. In England and Wales, the water and sewerage companies provide the economic regulator, Ofwat, with indicators of service performance covering water supply, sewerage services, customer service and environmental impact. Ofwat publishes the indicators annually in a public report. These simple performance scorecards have helped measure the efficiency of service provision and pressure the "worst in the class" (Kingdom and Jagannathan, 2001). Such tools need to be used with caution, however, as differences in performance can be due to a variety of factors aside from relative efficiency, such as differences in physical conditions, population density, nature of the terrain, age of the network, etc. For this reason, Ofwat has developed sophisticated econometric models to assess comparative efficiency of regulated companies while controlling for exogenous factors that can affect performance.

Opting for different levels of service to reduce initial capital costs

Choice of hardware and technologies can make a big difference to costs. In OECD countries, the regulatory regime in place can influence the selection of investment options, linked to the set of incentives that they introduce. While a rate-of-return regulatory regime may give an incentive to select higher cost options to earn a higher return (what is sometimes referred to as "gold plating"), incentive-based regulatory regimes (such as price cap regimes) introduce incentives to invest at least cost. In England and Wales, for example, this has allowed substantial investments to take place in the context of minimal tariff increases for customers. Optimising existing WSS infrastructure can generate substantial savings, for example, by scaling down capacity to the present and forecasted demand, or replacing inefficient pumps with a short asset life by new more efficient ones with a long asset life.

For many developing countries, particularly in Sub-Saharan Africa, the cost of reaching the MDGs appears far above currently available resources. The MDGs allow for a broad range of options to deliver improved water and sanitation, however, what is commonly referred to as a "service ladder".

At world level, the per capita costs of different options for meeting the water MDG have been estimated by Hutton and Bartram (2008): the report shows that the per capita cost of household connection is over three times higher than a stand post in Africa and Latin America. According to their estimates, the total global costs of attaining the water and sanitation MDGs could therefore go down from a high technology to a low technology option, from USD 327 billion to USD 135 billion, equivalent to an annual saving of USD 19 billion worldwide. Cutting down on investment costs may also be achieved by lowering service standards to levels that a country can afford: for example, many developing countries have adopted Western standards without tailoring them to their own circumstances, resulting in unnecessary investment costs.

3.2. Closing the gap: A combination of the 3Ts

As described in OECD (2009a), the 3Ts (defined as tariffs, taxes and transfers from overseas development assistance or philanthropic donations) are the ultimate sources of finance for water and sanitation services.[2] The 3Ts can also be used to leverage, and eventually repay or compensate, other funding sources, principally loans, bonds and equity, as discussed in section 3.3.

This section presents the concept of the 3Ts and discusses each of the Ts in turn. Each country is likely to adopt a different mix of the 3Ts to meet WSS's financing needs. Most countries have used public transfers (either from their own government or from external sources) to fund the development of WSS, particularly for capital expenditure. As countries develop and WSS become more mature, there tends to be a shift towards more use of commercial finance, reimbursed by growing cash flows from user charges (*i.e.* tariffs). For example, as set out in OECD (2009d), whereas tariffs represent 90% of direct financial flows to the sector in France, they only account for about 40% in Korea, 30% in Mozambique or as little as 10% in Egypt.

The mix of the 3Ts that is adopted by each government can have a substantial impact on the efficiency of the services. For example, in the US, switching from grant financing for capital investment (as used in the 1980s) to reliance on subsidised loans with long tenures and low interest rates (from the 1990s) brought significantly improved capital investment efficiency

(OECD 2009a). Therefore, OECD (2009a) underlined the importance of strategic financial planning to find the right mix of the 3Ts for achieving water and sanitation targets and leveraging repayable sources of finance.

3.2.1. Increasing revenues: Tariffs

Although the conventional economic wisdom calls for charging WSS tariffs at full cost recovery level, very few countries, either developed or developing, recover all costs via tariffs. This is true even when only financial costs are included and even more difficult when attempting to recover environmental and social costs. According to OECD (2009b), "sustainable cost recovery" (as originally defined by the Camdessus report) should be based on the simultaneous application of three principles:

- An appropriate mix of the 3Ts to finance recurrent and capital costs, and to leverage other forms of financing;
- Predictability of public subsidies to facilitate investment (planning),
- Tariff policies affordable to all, including the poorest, while ensuring the financial sustainability of service providers.

Tariff setting is usually driven by a combination of factors, many of which go beyond the immediate needs of the service. Politicians can insist on keeping tariffs low (*i.e.* below cost-covering levels) as water is an essential good, for which charging can be politically and socially sensitive. The "willingness-to-charge" may therefore be lower than the willingness-to-pay due to political motivations. From an economic perspective, setting tariffs needs to reconcile a series of potentially conflictive objectives, including economic efficiency, cost-recovery (or financial sustainability) and social concerns (or affordability). As discussed in OECD (2009d), a number of tariff structures can be adopted to reconcile those principles.

In OECD countries, operating costs are by and large covered but the scope for covering capital costs varies substantially. In OECD countries, OECD (2009d) found that prices can vary by a factor of 10 or more, ranging from 0.49 USD/m^3 in Mexico to 6.7 USD/m^3 in Denmark (such high price being underlined by an attempt to incorporate environmental costs into pricing). The report also sought to estimate cost-recovery ratios, based on IBNet data (see chapter 8) and other sources. Such analysis indicated that, in OECD countries, operation and maintenance costs of domestic and industrial WSS services are generally covered through tariffs. However, there does not appear to be a large margin for operators to also face the need to renew and replace ageing infrastructure, although very few countries provided data on this item. Generating revenues to cover the full economic or sustainability costs (including the environmental impact of abstracting water) seems to be a remote target only. An analysis of specific cases (such as Finland, Switzerland or Belgium) suggested that efforts have been made to increase cost-recovery in many OECD countries, and in particular to cover the costs of wastewater management where larger investments are needed.

Overall, WSS tariffs represent only a small share of average household incomes in OECD countries (ranging from 0.2% in Korea to 1.2% in Poland). These average figures hide some areas of "water poverty", however, with WSS bills representing up to 4.2% or 7.9% of household income for the poorest decile in Mexico and Poland respectively.

Cost-covering tariffs are much less prevalent in developing countries. OECD (2009d) indicated that prices for water supply and sanitation services in developing countries have been increasing over the last decade, however from usually low levels. Some countries in Asia, Latin America and the Middle-East have tariffs above 1 USD/m^3. However, in most cases, tariffs provide little incentives to use water efficiently (including by curbing down leakages) and do not cover costs. While operating costs are not always covered, capital expenditure for large investments is almost always financed via public funds, either from government taxes or international transfers (see sections 3.2.2 and 3.2.3).

In some regions, such as in Sub-Saharan Africa, households' contributions to sector financing are substantial, however, in the form of direct investments in self-supply. For example, the Africa Infrastructure Country Diagnostic, a continent-wide effort led by the World Bank to track expenditure in seven infrastructure sectors, found that households were actually the largest source of finance in the sector, ahead of domestic governments and international donors: "in Sub-Saharan Africa, households are important financiers of capital investment (0.3% of the Sub-Saharan African GDP) and account for USD2.2 billion, most of it dedicated to the construction of on-site sanitation facilities, such as latrines. The level of contributions from OECD donors is similar to that of domestic public resources and is equivalent to 0.2% of the Sub-Saharan African GDP" (Banerjee *et.al*, 2011).

In many developing countries, generating additional revenues via tariff reforms (including changes to tariff levels and tariff structures) requires taking account of affordability constraints for the most vulnerable population. According to OECD (2009a), the apparent trade-off between financial sustainability and affordability can be addressed via careful tariff design. Affordability can be assessed at two levels: for society as a whole, and for the most vulnerable groups (what can be referred to as "micro-affordability"). A number of countries (in the OECD and elsewhere) have adopted increasing block tariffs, with a first "subsistence" block provided at zero or very low prices. The assumption behind their adoption was that they would enable poor households to have access to a basic level of water services for free or at low cost, while at the same time contributing to cost recovery by providing a cross-subsidy from larger water users and providing an incentive to conserve water. But the actual experience with their implementation has shown that IBTs are regressive in countries with incomplete networks, where the poor are generally not connected and therefore do not benefit from the consumption subsidy by definition. Part of this results from the flawed design of IBTs in a number of countries (*e.g.* the lack of attention given to their impact on large poor households). Adjustments in their design can improve their capacity to target the intended population, but cannot completely overcome the shortcomings. In reality, poorer households are often larger households, so that they may end up consuming more than smaller, higher income ones. In areas where access is still low, it has been shown that the targeting performance of consumption-based subsidies is lower than that of connection subsidies (Komives *et al.*, 2005).

Alternative solutions to tackle affordability, apart from modifying tariff structures, include providing income support (to compensate poor households for increases in the prices of services of public interest that are judged to be unacceptably burdensome) and facilitating payment (to help poor consumers manager their budgets by paying water bills at short intervals for example).

In the context of the financial crisis, raising tariff revenues is likely to remain difficult. The financial crisis is likely to affect the ability for water companies to raise tariffs in two main ways: through a hardening of affordability constraints and a possible increased political reluctance to increase tariffs to sustainable cost recovery levels. The affordability constraint will be particularly felt in developing and transition countries. Although developing countries initially appeared to be shielded from the sudden stop in private capital flows that characterised the financial crisis from October 2008, they were later affected as the financial crisis spread to the real economy. In developed countries, household incomes are also stretched and consideration will need to be given for people on low income or with special needs who face increases in the cost of their utility bills and other costs in general.

3.2.2. Increasing revenues: Taxes

In both OECD and developing countries, allocations from public budgets still represent an important share of revenue for the WSS sector and are likely to play a significant role for the foreseeable future. According to OECD (2009a), the allocation of public funds to WSS can be justified for a number of reasons, including to promote the consumption of merit goods (whose value consumers may not fully realise, such as household sanitation and hygiene) or to compensate for market failures, by rewarding WSS providers for supplying public goods (public health) and external benefits (such as avoidance of groundwater pollution). Public funds may also be used to allow service providers to provide services at a tariff below cost for vulnerable consumer groups.

In order to be efficient and effective, subsidies should be transparent, targeted and ideally taper off over time. The most widespread form of subsidy among OECD and developing countries alike is capital expenditure. In OECD countries, for example, most of the heavy initial investment that was made in the late 19th and early 20th century (for water supply and sanitation) and since the 1960s (for wastewater treatment) were financed through public funds. Such capital expenditure subsidies can be provided in the form of grants, subsidised loans or guarantees, while utilities are expected to cover their O&M costs from tariffs. When utilities are owned by municipalities, local government budgets are often not sufficient and benefit from transfers from the central government. It is the case for example in South Africa, where municipalities often struggle to obtain adequate financing from tariffs. The central government therefore transfers municipal infrastructure grants, to address the capital investment backlogs inherited from the Apartheid era and the "equitable share", which is a need-based allocation transferred to local governments for operating expenses (Water Dialogues, 2009).

It is crucial that such transfers are provided in a way that ensures an effective contribution to the long-term sustainable financing of the WSS sector. Experience gained in the OECD and in countries of Central and Eastern Europe shows that two important criteria should be taken into account when organising these transfers: intergovernmental transfers should generate stable revenues that can be integrated in medium-term financial strategies of local governments and those transfers should be limited in time, until the achievement of pre-specified targets (EAP Task Force, 2006).

While public funds are limited by budgetary constraints and multiple demands from different sectors, there is scope for increasing public budget spending. In particular, several developing countries currently allocate only a small portion of government spending to the water and sanitation sector. Results from a recent survey of expenditure on water and sanitation, reported in the GLAAS report (UN-Water, 2010) state that countries reported public expenditures (from internal and external sources) between 0.04% and 2.8% of GDP for drinking water and between 0.01% and 0.46% of GDP for sanitation. Among the countries that had responded, Burkina Faso was the country that spent most on water and sanitation combined as a percentage of its GDP (with an estimated 3% of GDP), while countries with the lowest expenditure on the sector as a percentage of their GDP included South Soudan, Ivory Coast but also the Philippines. These figures highlight that overall spending in developing countries remains insufficient, although they mostly account for public spending and do not include private sources of finance. It is also recognised that data on national government spending on water and sanitation is not always very robust.

In OECD countries, in the context of the economic crisis, however, tax transfers are only likely to surge where stimulus packages target the water sector. The financial crisis is likely to have a two-pronged effect on government transfers to the water sector, as set out in OECD (2010a). A potentially negative impact is that, during times of crisis, there are many competing demands for limited public funds. Substantial public borrowing is likely to exacerbate the pressure on non-sovereign borrowers, through a "crowding-out" effect, making it even harder for them to borrow at acceptable rates. On the other hand, several governments have responded to the crisis by unveiling substantial stimulus packages, which could benefit the water sector. Following the lead of the United States and China, many of these stimulus packages include measures to "green the economy" (such as the "Green New Deal" announced in Korea) which, in some cases, include investments in water and wastewater.

In addition, governments in developed and developing countries alike are less likely to be able to borrow at acceptable rates. As a result, they may be tempted to make "temporary" cuts in water and wastewater investments so as to reallocate those resources to other sectors, with potentially long-term damaging impacts. The economic and financial crisis will also strengthen the case for making the best use of public resources (taxes and ODA alike) in order to leverage other forms of finance, including repayable finance (see section 3.3).

3.2.3. *Increasing transfers (*i.e. *Official Development Assistance and philanthropic donations)*

Official Development Assistance may be able to play a role in closing the financing gap in transition and developing countries. The share of ODA to water and sanitation varies across recipient countries. In some countries ODA subsidises most investments, while in others it plays a more marginal role. ODA has an important role to play both as a source of finance and of capacity development for the provision and financing of water services. It can also have a catalysing effect by reducing bottlenecks (particularly capacity constraints), ensuring access to the poor, and harmonising and aligning assistance with national strategies.

While the bulk of ODA is extended in the form of grants, loans constitute a large share of ODA to certain sectors. About half of ODA to water supply and sanitation in 2001-06 was in the form of loans. In the context of an analysis that distinguishes between the basic sources of revenue (tariffs, taxes and transfers) and other financial means, the different roles of ODA grants and loans need to be borne in mind. *ODA grants* consist of "transfers" and are considered as basic sources of revenue. *ODA loans* lower the cost of capital and are useful in helping water utilities "bridge" the financing gap that is created by the need for large upfront infrastructure investment and are therefore rather to be accounted for in the category "repayable sources of funding" (see next section).

After a temporary decline in the 1990s, **aid to water and sanitation has risen sharply since 2001**. In 2008-09, total annual average aid commitments to water and sanitation amounted to USD 8.1 billion. Bilateral aid to water increased at an average annual rate of 18% over the period 2002-09 and multilateral aid also rose by 10% annually.

According to OECD-DAC (2010), the share of aid to water and sanitation in DAC members' aid programmes has also risen since 2001, although at a more modest pace. In 2008-09, aid to water and sanitation represented 8% of DAC members' bilateral sector-allocable aid, as shown in Figure 3.2.

There are issues about ODA's current allocation, which means that targeting could be improved in order to deliver maximum benefits. Over the 2003-08 period, loans represented just over half of total aid to water and sanitation. Projects for "large systems" were predominant and accounted for

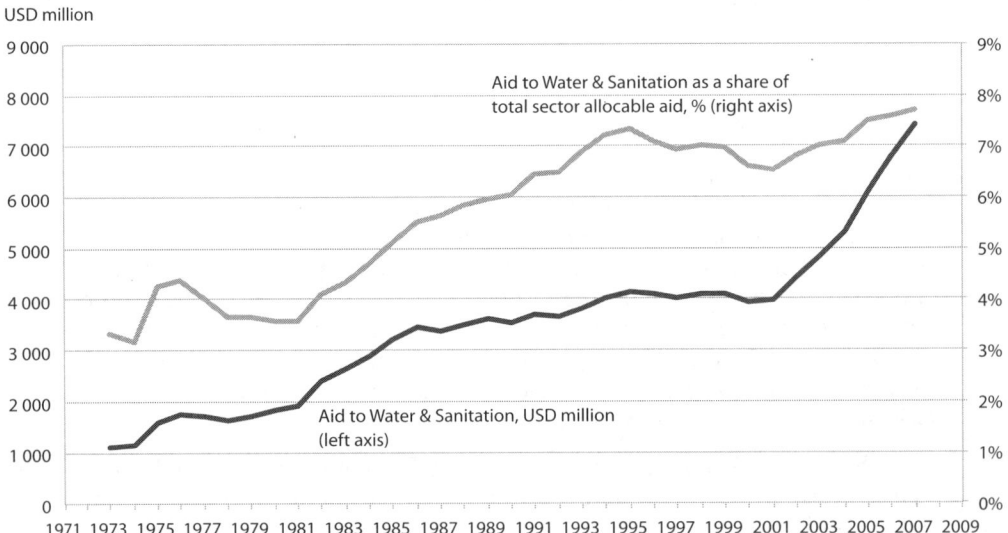

Figure 3.2. **Volume and share of aid to water and sanitation**
1971-2009, commitments, 5-year moving averages, constant 2008 prices

Source: OECD-DAC (2010), *Financing Water and Sanitation in Developing Countries: the Contribution of External Aid*, OECD, Paris.

57% of total contributions to the water and sanitation sector in 2007-08, with 68% of total ODA for large systems (as defined in the Creditor Reporting System database) in the form of loans. By contrast, donors relied almost exclusively on ODA grants (90% of total) to finance basic drinking water and on-site sanitation. During that period, aid to water and sanitation primarily targeted regions most in need of improved access to water and sanitation: Sub-Saharan Africa received 29% of total aid to the sector, and South and Central Asia 18%. Poorest countries classified as "low income" received 43% of total aid to the sector, two-thirds of which was in the form of grants. However, substantial levels of aid for water and sanitation go to middle-income countries where unserved populations are relatively low. By contrast, OECD (2009a) highlights that numerous countries with low levels of access receive comparatively little aid. Finally, aid tends to be allocated more to drinking water than to sanitation, although it remains difficult to assess this breakdown with precision. When development projects combine both water and sanitation, drinking water tends to receive the lion share.

International transfers will increasingly be needed to fill the gap but will be affected by the dire situation of public finance. In the context of the financial and economic crisis, it is likely that international transfers from IFIs, bilateral donors and charitable organisations will increasingly be needed to fill the financing gap in the water sector or to leverage other sources of finance, including market-based repayable finance. On the lending front, some IFIs have seen a growing demand for their services and products, especially as the competition from commercial banks has reduced. It should be noted, however, that such IFIs have to finance their loans through the capital markets and that their own borrowing costs have increased in line with the market.

A critical question is whether international donors (and philanthropic organisations) are going to be able to significantly increase their commitments in the years to come. Previous economic crises have usually seen official development assistance fall. Given the global significance of the crisis, however, OECD governments have committed to maintaining aid flows despite pressures on their own budgets. This has also been reflecting in IFIs expanding their lending facilities as a response to the crisis. For example, the World Bank has set up specific facilities to address what they identified as the sectors most at risk, with a special focus on infrastructure.

Concessional lending may fare better than grants in this context. For example, the AFD, the French bilateral donor, has been gradually reducing the proportion of grants to the water sector as opposed to loans. Development banks, on the other hand, have been increasingly called upon: for example, the EBRD in Eastern and Central Europe, where commercial lending for water has been drying up, has had a boom year in 2009. Difficulties in attracting market-based repayable finance to the sector and the potential need for innovation are discussed in the next section.

As aid resources are likely to remain relatively scarce in the foreseeable future, OECD (2009a) highlights the need for these resources to be spent strategically, so as to maximise their leveraging capacity and effectiveness. Areas where ODA can have a catalysing effect include reducing bottlenecks in the sector, supporting the financial planning process, ensuring access to services by the poor and supporting the development and use of risk-management mechanisms that can help attract private funding (and local private funding in particular).

3.3. Bridging the gap: Tapping repayable sources of funding

This section examines the role of repayable sources of finance in order to bridge the financial gap. While revenues from the 3Ts can *close* the financing gap for WSS, the role of repayable finance is only to *bridge* the financing gap, since it requires subsequent compensation in the form of interests or dividends. WSS providers usually look to mobilise repayable finance in order to finance capital expenditure for repairs, renewals or expansion of water and sanitation systems while ongoing operating costs and ordinary maintenance are routinely financed from a mix of the 3Ts (OECD, 2010a).

Market-based repayable finance refers to a sub-set of repayable finance, where financing is provided through the market by private actors. Sources of market-based repayable finance include: debt finance (loans from commercial banks, bonds issued through capital markets, project finance) and equity finance (from capital markets or private equity funds). Debt financing has been the backbone of most infrastructure investment in developed countries. Depending on the development of local bond markets and the size of the debtor, it has come either in the form of bonds or loans. In developing countries, water companies can use bank loans to finance capital investments (although these are usually concessional loans from development institutions). The use of other forms of finance, such as bond finance, project finance or equity finance has so far remained limited in developing countries but they are gradually emerging as ways to complement other forms of finance.

Below we examine the current extent of the use of repayable financing in the water sector (mostly in OECD countries and in a limited set of emerging countries) and how public financing could be best used to leverage such repayable financing so as to alleviate a number of constraints limiting their more extensive use in developing countries.

Bank finance. Short and medium-term commercial loans are common for financing working capital requirements in developed and developing countries alike. Short and medium term lending facilities may also need to be used to finance investments in countries where obtaining long-term bank financing to match the long asset life of water sector investments is difficult, as commercial banks are not able or willing to lend over such long periods. In developing countries, commercial banks are usually not familiar with the water sector, which is perceived as a high risk sector due to difficulties with increasing tariffs, inefficient management and corruption. Water utilities' revenues may not be sufficient to reimburse loans. Furthermore, they may not be sufficient to cover market-based financing costs, and this limits their ability to borrow. Finally, certain types of service providers, such as local or small-scale service providers may not have access to traditional bank financing at all, although they may have the option in some cases of relying on microfinance institutions for access to credit.

The re-evaluation of risk that has taken place during the financial crisis has led to a dramatic increase in the cost of commercial debt and in a reduction in the availability of overall debt financing, especially for long-term debts, resulting in a severe contraction in bank lending. The onset of the financial crisis has also affected sovereign states' ability to borrow and consequently reduced the value of sovereign guarantees in some cases. Microfinance institutions have suffered as well and may be less willing to diversify in water and sanitation away from their more traditional markets, *i.e.* income-generating activities. However, microfinance institutions in many developing countries are not offering such micro-loan/finance facilities for WSS. National development banks – if and where they exist – tend to focus more on large WSS projects than on small ones. As a result, bigger, richer and creditworthy cities usually can obtain bank finance, while most small towns and rural areas are neglected.

Bond finance. In developed countries, the water sector is considered to have a low risk profile that makes it well suited to the debt market. Bond financing is common in developed markets as it often offers cheaper access to debt finance than loans. The types of bonds issued can include corporate bonds, sovereign bonds or municipal bonds, depending on the structure and ownership of the water sector. For example, in the United Kingdom, the water market is dominated by large private water and sewerage companies which issue corporate bonds.

In the United States, water companies are smaller municipally owned companies and municipal bonds have provided a major source of finance for water and sanitation investments in the US since 1837. The financial crisis has affected such source of finance on the US market, however, as the credibility of credit rating agencies has been questioned and several monoline insurers (which used to enhance the rating of municipal borrowers in exchange for an insurance premium) have disappeared. As a result, highly rated municipal bonds have somewhat lost their attractiveness for cautious investors, making it difficult for US municipalities to raise the budgeted funds. In the majority of less developed markets, municipal bonds were not available even before the onset of the crisis due to poor creditworthiness and transparency of those entities. There are a few exceptions, with incipient municipal bond markets in India, the Philippines or South Africa which have been used partly to finance water and sanitation investments.

Project finance. Project finance consists of financing long-term infrastructure through a special purpose entity that can be financed with project debt and equity. A project finance "deal" would typically involve a number of equity investors, known as "sponsors" and a syndicate of banks that provide loans to the operation. Following the financial crisis, the feasibility of project finance deals based on high debt levels granted to off-balance sheet special vehicles has been severely affected, particularly in countries considered to be risky. New project finance structures are likely to require co-operation with sovereign-backed banks and will often require bridging loans at less favourable conditions.

Equity finance. Raising equity can be a good way of financing long-term investments as it is a source of finance with no specific deadline for repayment. Equity holders are usually interested in holding their stake over the long term in order to benefit from future dividends and any potential increase in the value of their equity. Equity can be used as collateral to leverage other forms of private finance, rather than as a way to finance long-term capital investments directly. When equity investors are private, however, that would usually be reflected in a higher cost of equity versus the cost of debt finance. Shares are either listed on a stock exchange (which can be referred to as the "listed equity model") or held privately, by the founders and managers of the company or institutional investors. A number of water companies have listed shares on the stock exchange, including some public companies (such as SABESP in Brazil) and private ones (such as Lydec in Morocco or Manila Water in the Philippines). However, a key constraint weighing on the ability to raise capital on the stock exchange is linked to the varying degree of development of local capital markets.

In the context of the financial crisis, equity financing has been more difficult to attract as the equity risk premium (*i.e.* the return expected by equity investors compared to risk-free investments) has gone up in both developed and developing countries.

The financial crisis has substantially affected listed water companies: for example, a weighted index of Asian water stocks was down 47.5% at the end of 2008 compared to its January 2008 value. American water stocks lost 5% of their value during the same period while European water stocks were down between 30 and 90% throughout 2008. For example, market leader Veolia lost 64% of its value during 2008 after issuing two profit warnings.

Among others, water American Water Works (USA), Nova Cerae (Brazil) and Maynilad (Philippines) had to postpone Initial Public Offerings (IPOs) due to the adverse market conditions.

In early 2009, equity valuations had bounced back substantially however, especially in Asia where the resurgence of water stocks was supported by renewed access to capital. The new environment could encourage more companies to recycle capital by spinning out business trusts. The GWI Water Index, which tracks major water stocks around the globe, was up 7.7% in May 2009. One year after, steady rise in all segments had taken the benchmark to its highest level since August 2008, with top performers in Asia.[3]

On the whole, however, availability of market-based repayable finance has been negatively affected by the financial crisis and the potential to rely on certain financial innovations seriously dented. This trend has to be placed in the broader context of the overall availability of finance to the sector, however, so as to assess the likely impact on investments going forward.

In developing countries, a number of critical mismatches have limited flows of repayable finance for WSS. Market-based repayable finance is more difficult to mobilise for WSS due to a number of constraints, which OECD (2010a) referred to as "critical mismatches".

The sector is often perceived by potential providers of market-based repayable finance (such as banks, institutional investors, private equity funds, equity investors, project sponsors, etc.) as a "high risk/low return" sector, even though its fundamental economics (with relatively stable and almost "recession-proof" demand for the services and long-life buried assets) would rather place it in the "low risk/low and steady return" category for a number of reasons. This high-risk reputation is frequently linked to difficulties for increasing water and sanitation tariffs to cover costs, due to perceived affordability constraints or political resistance to increasing tariffs. As a result, many water utilities are in dire financial situations, with under-capitalised balance sheets that impede their capacity to raise debts. In the absence of any repayment capacity or history of past lending, most commercial banks are unlikely to lend to the sector which they do not perceive as being "creditworthy".

In addition, local financial markets may not be able to provide long-term loans with low interest rates to water operators, which overwhelmingly tend to be mid-size of small utilities, which can be referred to as an inappropriate "market fit". There is often a discrepancy between long-term investments needed in the water sector to match the life of the assets and the short-term lending capabilities on local markets. Informal operators, who serve an average of 50% of the population in developing countries according to Kariuki and Schwartz (2005) have difficulties in accessing credit from the conventional banking sector. In many countries, decentralisation of water and sanitation services has transferred large investment needs to local government and utilities. However, the availability of funds at local level is restricted: local government's creditworthiness tend to be low, making it challenging to raise funds on international markets, and the small scale of service of many utilities may result in too high transaction costs to make market-based financing viable.

Finally, as highlighted in OECD (2009c),[4] WSS combine a number of substantial risks, such as commercial risk (related to revenue), contractual risk, and foreign exchange risk that make equity capital and debt financing from international markets expensive and may deter commercial funding.

Innovative financing can play a major role to attract market-based repayable finance to the sector. Financial innovation could significantly help with leveraging market-based repayable finance into the water sector, both in OECD and developing countries.

Table 3.1 outlines examples of critical mismatches in the sector and the types of innovative financial mechanisms that can be used to address those constraints. In the context of some of these innovations, public funds could be used in order to leverage market-based repayable financing so as to increase the overall amounts of finance available to the sector.

Below are a few examples of what these innovations might entail; OECD (2010a) contains more detailed analysis of these innovations and examples of where they have been applied.

Blending grants and repayable financing consists of combining concessional financing (either grants or loans with a grant element) with repayable finance in order to support a single project or a comprehensive lending program. In the water sector, this has been done at the level of specific projects, like in Maputo (Mozambique) for the financing of the urban water and sanitation program or via the establishment of financing vehicles, which aim to combine diverse sources of finance (such as in FINDETER in Colombia, a public-private financing entity which rediscounts commercial bank loans for local infrastructure development, including water and sanitation). Such blending can take many forms: ODA grants can be provided as interest rate subsidies, seed financing for revolving funds or contributions to the establishment of project preparation facilities. The main objectives of blending are to attract funds that would otherwise not be attracted by a given project while ensuring that basic public policy goals, such as increasing access and serving the poor, are met. Such structures hold great potential in the water sector, especially in the context of the financial crisis, given that an element of subsidy is almost always going to be required to make a water sector project bankable and reach the underserved at the same time.

Table 3.1. **Examples of innovative financial mechanisms in the water sector**

Critical mismatch	Examples of innovative financial mechanisms
Affordability constraints at household level	• Blending grants and repayable financing • Micro-finance • Output-based aid
Limited availability of funds for domestic operators and SSWSPs	• Micro-finance • Output-based aid and innovative contract
Risk profile and difficulties in managing certain risks (*e.g.* political risk, foreign exchange risk)	• Blending grants and repayable financing • Guarantees and risk insurance • Devaluation backstopping facility • Local-currency financing • Revenue agreements in lieu of guarantees
Lack of funds at decentralised level	• Municipal bonds • Pooled funds, revolving funds and bond banks • Instruments to increase sub-sovereign lending
Short tenor of available financing	• Guarantees • Equity contributions
Under-capitalised balance sheets	• Raising equity to strengthen the balance sheet, convertible loans, debt-equity swaps, "asset-light" expansion models
Lack of understanding by external lenders and investors	• Blending grants and repayable financing • Credit ratings • Project preparation facilities
Lack of "bankable" projects	• Project preparation facilities

Source: OECD (2010), *Innovative Finance Mechanisms for the Water Sector*, OECD, Paris.

Microfinance has been identified as a key way to overcome affordability constraints for providing access to services, particularly for households and small-scale water providers in developing countries. The use of microfinance has so far been limited in the water sector, partly due to a lack of awareness and limited understanding on the part of microfinance and water sector professionals of their respective sectors. However, a recent review by Mehta (2008) made the case for the strong potential of microfinance in the sector, particularly for loans to households and to community projects (such as slum redevelopment projects). ODA can play a role in developing the use of microfinance for WSS by providing seed financing to revolving funds or microfinance institutions, smart subsidies for product development or guarantees. Donors and IFIs can help build awareness of microfinance products, through capacity building activities or blending microfinance with other types of financing instruments in the projects they choose to support. For example, they can combine reliance on microfinance (or local commercial banking in the case of small-scale entrepreneurs) with the use of Output-Based Aid, *i.e.* subsidies paid based on effective and measurable results to service providers, which are therefore better incentivised to deliver results. Although a growing number of pilot projects have adopted OBA principles in the water and sanitation sector, the approach has yet to be mainstreamed. Increasing the use of OBA may require being more explicit about the need for pre financing, which could be achieved by combining OBA subsidies with access to microfinance, as it was done successfully in a landmark operation in Kenya with a local commercial bank (K-Rep). To reduce transaction costs over the long-term, setting up OBA facilities at country level could also be explored further so that project and service provider selection as well as contract monitoring can be carried out in-country rather than through an international institution.

Although a whole array of **guarantees and insurance products** are available from donors, IFIs and private institutions, they have not been used on a regular basis or at a large scale in the water sector. This partly reflects the changing structure of the market for water services: while international private operators have largely been driven away by adverse conditions, guarantees provided by international institutions for relatively large "transactions" are less appropriate than previously. Besides, IFIs and donors have usually maintained fairly rigid rules about the use of these guarantees (for example, with counter-guarantee requirements or restrictions on the provision of stand-alone guarantees), which means that transaction costs for applicants remain excessively high. The establishment of "guarantee facilities" at national level, to which donors and IFIs can contribute seed financing or overall guarantees (as done with LGUGC in the Philippines) could facilitate the provision of guarantees at the local level, which is more in line with the current market structure in the water sector. Donors and IFIs may also need to step in where private entities or governments have become less willing to provide guarantees.

Forming **grouped financing vehicles** can be a helpful way to provide access to finance to a large number of relatively small borrowers, particularly with the combined use of guarantees to improve credit rating. Such groupings are particularly well-suited to decentralised water sectors, in which small and medium-sized service providers are struggling to access financing on their own merit. In the sector, they have mostly been used as a basis for issuing bonds in countries with fairly mature financial markets, such as in the United States but also in Mexico or in India. High transaction costs and limited knowledge, once again, can partly explain why their spread has remained somewhat limited beyond those markets. Donors and IFIs would need to step up their efforts in order to create such structures or help define institutional environments that would be conducive for grouped financing vehicles to be established where appropriate. This may require establishing such grouped financing structures directly (such as revolving funds, bond banks, etc.) or fostering the adoption of legislation that make such structures more attractive (such as tax-exemptions on bonds issued by such structures, as practised in the US, or requirements that grouped financing vehicles be formed in order to access government financing).

Direct lending to sub-sovereigns, without the need for a central government guarantee has been practised with success for some time by some IFIs and donors, such as the EBRD or the AFD. However, many other donors and IFIs have not been able to lend at the sub-sovereign level, either because their internal rules do not allow them to do so or because they are not willing to take on a risk that they cannot manage adequately. Besides, sub-sovereign entities in many countries are either too weak financially to borrow or lack the capacity to put together a bankable project eligible for donor financing. Central governments themselves may not be willing to let sub-sovereign governments borrow directly, particularly when they are not able to keep control over the overall debt burden that is being accumulated at the national level (which they may have to cover ultimately in the event of bankruptcy, even if they have not provided an explicit guarantee). Donors should evaluate how they can relax such guarantee requirements at the sub-sovereign level, so as to pave the way for commercial lending to those borrowers. For example, reliance on revenue agreements with the sub-sovereign borrowers to either increase tariffs or intercept central government transfers can provide security to lenders without the need for central government guarantees. These types of agreements can help introduce financial discipline and support the implementation of reforms at the level of borrowers, as long as donors and IFIs can also provide adequate resources to support reform processes at the local level. Lending in local currency can also be a key tool to make such loans more attractive to local governments and water utilities. Finally, donors can combine these lending instruments with guarantees to commercial lenders so as to broaden the pool of financiers and investors interested in investing in water and sanitation at the local level. Direct lending to entities at the sub-sovereign level, such as municipalities or municipal utilities, can help those borrowers build a credit history and give them access to a broader range of investors, including commercial banks and equity investors.

Raising equity can help strengthen the balance sheets of water companies, which are often under-capitalised. Interesting models have been developed in the water sector to mobilise equity via financial markets (such as the Hyflux Water Trust in Singapore), thereby diversifying away from mobilising funds from private water companies (whose ability to bring in equity capital is limited in any case) and using such equity injections to leverage other forms of finance for capital investments. Mobilising equity through capital markets can strengthen financial discipline and improve transparency, including for companies that are primarily government-owned (including a number of State Water Companies in Brazil, which are publicly listed). When requested to provide equity in a distressed situation, many donors tend to be reluctant to do so as such equity contributions can sometimes be treated as an implicit subsidy when return expectations are very low. However, as long as financial discipline is maintained, equity contributions can strengthen the balance sheet and provide a sound basis for leveraging additional forms of finance, such as loans and bonds. In such selected cases, IFIs and donors can make such equity injections themselves, including in some cases by swapping debt for equity.

Credit ratings can help improve transparency and facilitate access to financial markets for borrowers. Significant progress has been made for awarding credit ratings to municipal governments and water companies, although the use of such ratings has remained limited, particularly in markets that are too small to develop a national rating scale. The financial crisis has significantly affected the credibility of rating agencies, however. Donors and IFIs can potentially step in to develop "shadow" credit rating systems *i.e.* based on a one-off exercise, to examine the creditworthiness of particular companies and make recommendations on how they could improve their creditworthiness. Other donor initiatives to improve overall transparency and improve knowledge of the sector for external financiers should also be encouraged. For example, the benchmarking system IBNet set up by the World Bank could be strengthened so as to improve the reliability and comparability of the information produced. Overall, a lot of information is already collected by different institutions. Donor-led efforts to improve its quality, increase co-ordination between sources and disseminate its existence could have a positive impact on raising the profile of the sector with external financiers.

Finally, ***project preparation facilities*** can also help with the definition and preparation of bankable water projects. A limited number of such facilities have been set up at the international level. Project preparation facilities, on the whole, have enabled the preparation of bankable projects in an accelerated manner and improved the effectiveness of donors' contribution by pooling funds together for support to project preparation. They have been particularly useful in well-defined geographical areas where they have been set up to accompany well-defined policies, such as in Eastern Europe or the Mediterranean. In Sub-Saharan Africa, they can be particularly useful to assist countries with limited project preparation capacities to develop projects that can only attract repayable finance if they are combined with innovative approaches to financing, such as blending grants and loans or using guarantees to reduce the risk perception. In the future, donors and international organisations can help finance the establishment of such facilities so as to prepare projects that they are either willing to finance themselves or to attract market-based repayable financing to (provided projects prepared in such a way can receive funding from a diversity of sources). The establishment of such facilities at the national level could also be encouraged, as it can reduce transaction costs and tie more easily into domestic financial mechanisms outlined earlier. Beyond the setting-up of project preparation facilities, local expertise for project preparation should be strengthened, from project conceptualisation all the way down to design stage and implementation.

3.4. Mobilising the private sector

The private sector is involved in many different ways in the water sector. As mentioned in OECD (2009a), private actors alongside the different segments of water service provision may include:[5]

- *Formal private water and sanitation service operators.* They provide services based on a contract or license with the public authorities responsible for delivering services in a given region or country.[6] They may be either domestic operators, which tend to focus on a single country, or regional and international operators, which provide services to customers in a broad range of countries.

- *Informal private water and sanitation service operators.* Small scale independent providers generally operate informally where no public services are provided or provided at low standards of quality;[7] large-scale developers of both residential and commercial property often provide and operate water and sanitation services to their own developments thereby making a considerable contribution to a city's infrastructure and service coverage.

- *Private financial institutions (such as banks or investment funds).* They provide financing from private sources, either through commercial bank loans, bonds or capital in the form of equity. Such private financing can be provided to both public and private operators;

- *Private companies* whose main business is not water but which are heavy water users (such as beverage, mining or construction companies).

This section focuses more specifically on the role that formal private operators can play in mobilising financing for the sector. During the 1990s and early 2000s, the introduction of private sector participation (PSP) in the management of water and sanitation services in developing countries was somewhat wrongly construed as a way to bring additional financial resources to the sector and therefore to fill the financing gap. Indeed, the introduction of private sector participation (PSP) was often based on the misconception that private operators would bring financing with them in the context of concession contracts or other similar contracts with investment obligations. The early termination of a number of high-profile concessions (such as in Buenos Aires) following financial crises, in which the private operator was exposed to foreign exchange risk on its debt to finance investment programs, challenged these earlier expectations. It also helped highlight the fact that private operators themselves have to source external capital and arrange financing.

More generally, recent experience has allowed to gain a better understanding of the ways in which private operators can either directly or indirectly mobilise financing for the sector, which they can do:

- ***By improving overall sector efficiency, thereby reducing costs (and financial needs) and improving the sector's creditworthiness and ability to attract financing.*** By reducing costs private sector participation can contribute to fill (*i.e.* reduce) the financing gap. Improved services can contribute to creating a "virtuous circle": customers are more willing to pay their bills when service improves, more efficient operation increases cash flow from operations, more funds are available for investment, which in turn increases the customer base and the utility's revenues. As creditworthiness improves, a utility can more easily access funding and invest in service expansion;

- ***By financing investment costs, particularly when the public sector's ability to borrow is limited.*** Private operators are sometimes brought in because they are deemed more able to mobilise financing, especially from private financial institutions. While the facilitation of access to repayable market finance is a crucial role that PSP can play, especially given the need for such funding to cope with huge upfront capital investment costs, it does not per se contribute to fill the financing gap, but rather helps to bridge it, private financing will ultimately need to be repaid (plus interest) through a combination of the 3Ts.

- ***By managing and enabling the capital programmes of public authorities.*** The private sector manages an extensive investment programme on the behalf of the public authority and co-ordinates the work with the ongoing operation and maintenance of the service. This has made a significant contribution to increasing public sector investment into the sector in cases such as Algiers.

The first of these points is supported by empirical evidence. Private sector participation over the last 20 years has proved to be a useful tool for improving sector performance and efficiency, as shown by a number of recent studies. For example, Marin (2009) looked back at 15 years of experience with public-private partnerships (PPPs) for urban utilities in developing countries and evaluated their impact on four dimensions of performance: access (coverage expansion), quality of service, operational efficiency and tariff levels. Marin's research found that many private operators succeeded in reducing water losses, notably in Western Africa, Brazil, Colombia, Morocco and Eastern Manila in the Philippines. In some cases, private operators reduced non-revenue water (NRW) to less than 15%, a rate similar to the best-performing utilities in developed countries. Such efficiency gains have contributed to improving the financial position of water utilities (by cutting costs, increasing revenues and therefore reducing the need for external subsidies) and to reducing (*i.e.* filling) the financing gap.

The second point, *i.e.* the ability of private sector participation to facilitate access to repayable market finance requires a longer discussion. The ability for private actors to mobilise financing largely depends on the type of contractual arrangements they have entered into with the public sector. Private companies can operate under a broad variety of contractual arrangements with the public sector, which reflect the ways in which risks have been allocated between the parties. The allocation of responsibilities for investment (and for financing such investment) can vary substantially according to the contractual arrangement in place.

An overview of the range of contractual arrangements is presented in Table 3.2. It is only in the case of concession contracts, BOTs, divestitures or some joint ventures that private operators are requested to mobilise substantial funding for capital investments directly, which they can usually recoup via tariff revenues or fees. In the case of other contractual arrangements, responsibility for mobilising investments rests with the public sector and the private operator is brought in largely for its capacity to drive efficiency gains or to mobilise financing indirectly. The private operator has also more control over management in the case of "higher-powered" forms of private sector participation (such as concessions, BOTs or divestiture), which is usually associated with a greater ability to deliver efficiency gains.

Overall, Marin (2009) notes that private financing of urban water utilities (*i.e.* new capital brought in by private operators) in developing countries has been limited when compared with other infrastructure sectors, as it represented only 5.4% of the total investment commitments in private infrastructure between 1990 and 2000. Based on figures from the PPI (Public-Private Infrastructure database),[8] he finds that investment commitments by private operators (made in the year of financial closure) have gone down sharply in the aftermath of the Asian financial crisis, from a peak of USD 10 billion in 1997 to a low of about USD 1.5 billion in 2003, and have not recovered since, as shown in Figure 3.3.

Another study (Gassner *et al.*, 2008) examined private investments based on detailed regression analysis of water and electricity PSP contracts

Table 3.2. **Typology of contractual arrangements between government (G) and the private sector (P)**

	Service contract	Management contract	Affermage/ Lease	Concession	BOT	Joint venture	Divestiture
Asset ownership	G	G	G	G	P/G	G/P	P
Capital investment	G	G	G	P	P	G/P	P
Commercial risk	G	G	Shared	P	P	G/P	P
Operations/ Maintenance*	G/P	P	P	P	P	G/P	P
Contract duration	1-2 yrs	3-5 yrs	8-15 yrs	25-30 yrs	20-30 yrs	Infinite	Infinite
Source of retribution of operator	Municipality	Municipality: fee is fixed or based on performance.	Operator collects user fees. Lease: fee paid by municipality Affermage: revenue shared	Users	Municipality	Users	Users
Occurrence 1991-2009 (World Bank PPI Database)	Not part of scope	Together: 111 of 715 projects		278 of 715 projects	294 of 715 projects	Not a separate category	32 of 715 projects
Examples	Mexico City Chennai	Johannesburg Amman	Cartagena Côte d'Ivoire Senegal	Gabon Jakarta Manilla	China India Malaysia Mexico Morocco	Cartagena Netherlands Chongqing Sino-French Water Supply	England Chile

* Maintenance may lead to considerable amounts of investments on the part of the responsible partner.

Source: OECD (2009), *Managing Water for All: An OECD Perspective on Pricing and Financing*, OECD, Paris. Updated based on World Bank PPI database.

Figure 3.3. **Evolution of investment in public private partnerships projects in developing countries, 1991-2009**

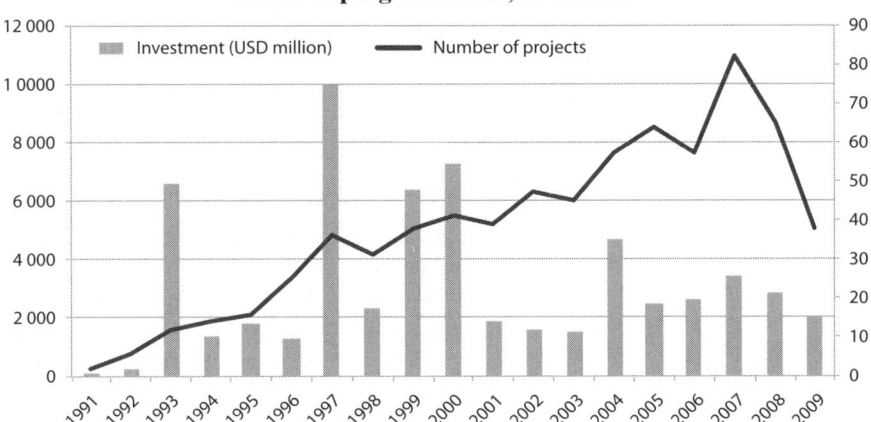

Source: World Bank Private Participation in Infrastructure (PPI) Database, *http://ppi.worldbank.org*.

(with a data set of more than 1 200 utilities in 71 developing and transition economies). The study found that it was not possible to conclude that investment always increases with PSP (despite evidence of actual increases in water connections, the number of which increased by about 12% on average).[9] For concession contracts, the study found no conclusive evidence that investment increased in developing countries. For lease and management contracts (where there is no investment obligation for the private party), the study suggested that the public asset holding company usually did not increase investment even if the PSP brought operational improvements.

This type of evidence has led more and more countries to adopt PPP models and risk sharing arrangements in which investment largely remains in the hands of the public sector while the private operator focuses on improving service and operational efficiency (OECD, 2009c and Marin, 2009).[10] In practice, funding for investment under these mixed-financing PPP projects comes from a combination of direct cash flows from revenues, with a variable mix of government and private sources that tend to make the traditional dichotomy between leases-affermages and concessions increasingly obsolete.

Several alternative approaches to combine private sector participation (so as to benefit from efficiency gains) and a mix of public and private financing have been developed over the past decade:

- Concessions that rely largely on revenue cash flow for investment, with cross-subsidies from electricity sales (Gabon), tariff surcharges (Ivory Coast), or both (Morocco).

- Affermages, as originally applied in Western Africa, bolstered by enhanced incentives for operational efficiency, a program of subsidised connections to expand coverage for the poor, and a gradual move to full cost recovery through tariffs (Senegal, Niger, and now Cameroon).

- Mixed-ownership companies, as used in Latin America (Colombia, La Havana in Cuba, and Saltillo in Mexico) and several countries of Eastern Europe (the Czech Republic and Hungary).

- Concessions with public grants for investments to spearhead access expansion or rehabilitation while minimising the impact on tariffs. This is typified by the PPPs in Colombia designed under that country's Programa de Modernización de Empresas (PME); a similar approach has been adopted in Guayaquil in Ecuador and in a few concessions in Argentina (Cordoba and Salta).

Finally, it is important to note that additional financial flows from the private sector at large (such as from households themselves investing in wells or latrines, property developers, commercial banks or bond investors) are not tracked in a comprehensive fashion in the water and sanitation sector, even though there is evidence that they represent a substantial portion of total investments (especially financial flows coming from household themselves). Market-based repayable finance can be provided to either public or private operators: for example, municipal bonds subscribed by private investors in the United States have largely financed municipal (and hence) public operators. This type of financing has therefore the potential to bridge the financing gap much beyond the limited universe of privately operated water service providers. In many countries, particularly developing ones, attracting such type of finance is likely to require financial innovation and public funds to unlock supply (and demand) of capital.

In sum, the private sector can contribute to filling the financial gap in several distinct ways. However, as developed in details in OECD (2009c), the ability of the private sector to contribute requires appropriate institutional and regulatory environments as well as sustainable cost-recovery. In order to provide guidance to countries on the allocation of roles, risks and responsibilities between public and private partners, as well as on the institutional, regulatory and policy framework necessary to improve the investment conditions in the water sector, the OECD has developed a tool – the *Checklist for Public Action* (OECD 2009c) – and supported its use in a number of countries, including Egypt, Lebanon, Mexico, Russia and Tunisia. The key lessons learnt from these experiences can be found in chapter 10.

3.5. Using strategic financial planning

The extent to which each source can generate additional funds will be highly location-specific and depend on the overall environment and on the willingness of governments to set realistic objectives and to adopt reforms so as to improve the efficiency and creditworthiness of existing service providers. As stated in OECD (2009a), "goals that are set politically and are not matched by real revenue streams result in major financing gaps and unexecuted plans, with the consequence that the poor suffer most through absent or deficient services". For example, Ethiopia has adopted a Universal Access Programme, which foresees improving access to improved drinking water sources from 22% in 2006 to 98% in 2012, but it is unclear how this policy would be financed. In some cases, donors share responsibility for lack of realism, for instance when they require the use of best available wastewater treatment technologies that may not be affordable if scaled up beyond the project level. Strategic financial planning must be carried out in the context of broader sector planning that address roles and responsibilities of government agencies, policy priorities and related legislative and regulatory reforms in order to ensure that a package of measures that can realistically be financed is being put forward.

In order to deal with those challenges, governments have to set realistic objectives for the development of the WSS sector, checked against available resources, and agreed in a multi-stakeholder policy dialogue (a process termed "strategic financial planning, or SFP"). According to OECD (2009b), SFP has several objectives: "it provides a structure for a policy dialogue to take place, involving all relevant stakeholders including Ministries of Finance, with the aim of producing a consensus on a feasible future WSS. It illustrates the impact of different objectives and targets in a long term perspective, linking sector policies, programmes and projects. It also serves the important aim of facilitating external financing, providing clear and transparent data on financing requirements". Such process can be carried out either at national level or at municipal or regional levels.

The OECD has supported the application of such approaches, using a strategic financial planning tool called FEASIBLE, in a number of countries, particularly in the EECCA countries such as Armenia, Bulgaria, Moldova, Georgia, Kazakhstan, the Kyrgyz Republic and six Russian provinces as well as in Egypt, Lesotho and Turkey (see chapter 5). OECD (2009b) drew key lessons from carrying out SFP in those countries, particularly in terms of conducting the process and deriving implications for reform.

For example, using the FEASIBLE planning tool, the OECD assisted the Government of Moldova with defining policy goals that they could afford (see chapter 5 and OECD 2009a for more details). Moldova faced a

Figure 3.4. **Annual cash flow needs and available financial resources in Moldova's water supply and sanitation sector (2006)**

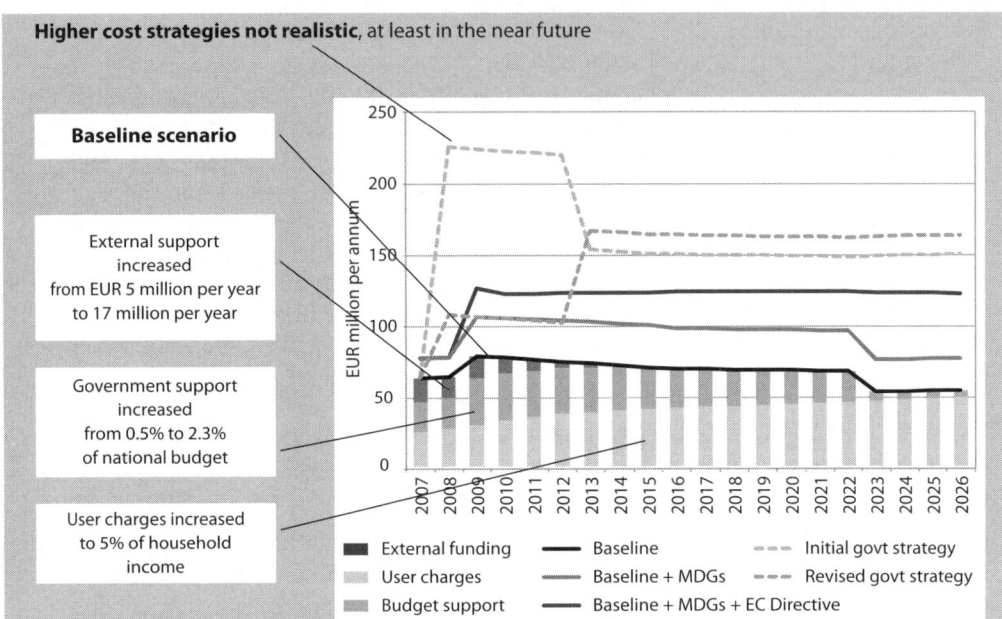

Source: OECD (2009), *Strategic Financial Planning for Water Supply and Sanitation. A Report from the OECD Task Team on Sustainable Financing to Ensure Affordable Access to Water Supply and Sanitation*, OECD, Paris.

situation where it still had to extend access in unserved rural areas while upgrading its existing installations and investing in wastewater treatment to meet European directives. The Government defined alternative investment strategies, ranging from a "baseline scenario" (which essentially assumed the maintenance and rehabilitation of existing WSS infrastructure, with no extension of service to previously not connected populations) to higher cost strategies including full compliance with EU directives. As shown in Figure 3.4, the analysis found that even if tariffs were increased substantially up to an average of 5% of household income (with social protection measures to support the poorest who would pay more than this average), user charges would only generate about 50% of cash flow needs for the foreseeable future, going up to 95% by 2028. The analysis therefore recommended prioritising investments, starting with investments to reduce water related morbidity and halt the deterioration of existing infrastructure, then improve the efficiency and reliability of existing systems. It was deemed necessary to achieve those goals first before considering extending existing systems (to meet the MDGs) or reduce pollution. This incremental approach is much more realistic and sustainable than a sudden increase in coverage that most economies cannot maintain in practice.

Countries where most benefits are to be reaped, *i.e.* where the access gap is the largest, are also the ones where the financing gap is the most glaring and will be most difficult to fill/bridge. For example, an evaluation of the financing gap to meet the MDGs and potential ways to bridge was recently conducted in the context of the Africa Infrastructure Country Diagnostic (AICD), a multi-donor initiative led by the World Bank. This evaluation showed that whereas middle-income countries in SSA may be able to reduce the financing gap to almost nothing thanks to performance improvements, the financing gap was likely to remain at a very substantial level in fragile states (see Box 3.1).

Box 3.1. **Evaluating the financing gap in Sub-Saharan Africa: The Africa Infrastructure Country Diagnostic**

A recent comprehensive review on the state of infrastructure in Sub-Saharan Africa was carried out by the Africa Infrastructure Country Diagnostic project, a multi-donor initiative led by the World Bank. For the water and sanitation sector, the study evaluated the financing gap to reach the MDG target and how such gap could be filled from existing or future sources.

Estimating current spending. The report found that existing spending on water supply and sanitation in Sub-Saharan Arica is USD 7.9 billion. The report found that household contribution to on-site sanitation facilities was higher than public spending either from public budget or ODA sources (0.3% of GDP spent by households on building latrines every year as opposed to 0.2% allocated by governments and 0.2% coming from ODA respectively). As such, they found that households contributed to almost half of total capital investments in the sector. Contributions from private sector operators were found to be negligible, with local capital markets contributing next to nothing to the WSS sector in Sub-Saharan Africa and little prospect for doing more.

The cost of reaching the MDGs. The report estimated that the price tag for reaching the MDGs for both water and sanitation in Sub-Saharan Africa would reach USD 22.6 billion per year, or 3.5% of these countries GDP. For improved water alone, it would be USD 17 billion a year (roughly 2.7% of SSA's GDP). Given the substantial access gap remaining in SSA, AICD (2010) estimated that capital investment needs for new infrastructure and rehabilitation of existing ones would account for over two thirds of total investment needs in some countries.

> **Box 3.1. Evaluating the financing gap in Sub-Saharan Africa: The Africa Infrastructure Country Diagnostic** *(continued)*
>
> ***Where will the money come from?***
>
> *The report then sought to estimate how the financing gap could be reduced, from a variety of sources, including the elimination of inefficiencies.* Table 3.3. shows the results of this evaluation.
>
> Table 3.3. **Funding gap (USD million per year)**
>
> | | Total needs | Spending traced to needs | Gain from eliminating inefficiencies | Sources of inefficiency ||| (Funding gap) or surplus |
> | --- | --- | --- | --- | --- | --- | --- | --- |
> | | | | | Under-execution of budget | Operating inefficiencies | Under-pricing | |
> | Sub-Saharan Africa | -22 640 | 7 890 | 2 877 | 168 | 1 259 | 1 450 | -11 873 |
> | Low-income, fragile | -4 531 | 441 | 471 | 6 | 106 | 358 | -3 620 |
> | Low-income, nonfragile | -7 810 | 1 840 | 685 | 39 | 265 | 381 | -5 285 |
> | Middle-income | -3 987 | 2 637 | 1 037 | 8 | 492 | 537 | -312 |
> | Resource-rich | -6 364 | 1 753 | 522 | 137 | 172 | 214 | -4 089 |
>
> *Source:* Ghosh Banerjee, S. and E.Morella (2010), *Africa's Water and Sanitation Infrastructure: Access, Affordability and Alternatives*, Directions in Development Series, The World Bank, Washington, DC.
>
> For example, the report estimated that losses associated with tariffs set below cost-recovery levels amounted to USD 2.7 billion a year in Sub-Saharan Africa and impeded service expansion. Improving cost recovery of water utilities could reduce the gap by USD 1.4 billion a year, and addressing operating inefficiencies would bring an additional USD 1.2 billion a year.
>
> However, the report concluded that even if major sources of inefficiencies were eliminated, the remaining funding gap would still be large, particularly in low-income countries. The report estimated that there was limited scope for increasing existing sources of finance, particularly domestic public finance and self-financing by households, which were both likely to be affected by the ongoing economic and financial crisis. They concluded that two realistic options to meet the targets would be to either defer the attainment of the infrastructure targets or to try and achieve them by using lower-cost technologies.

Box 3.1. **Evaluating the financing gap in Sub-Saharan Africa: The Africa Infrastructure Country Diagnostic** *(continued)*

The AICD report shows a wide range of fiscal efforts on water supply and sanitation throughout Africa. If the average is close to 0.9% of GDP, several countries find it possible to spend more than 2% of GDP on the sector. This leaves room for potential improvement in the other countries.

Source : Ghosh Banerjee and Morella (2010). See also *www.infrastructureafrica.org/aicd/*for more information on the overall Africa Infrastructure Country Diagnostic (AICD) project.

Where the financing gap remains substantial, public funding (in the form of domestic government funding or ODA) could potentially play a critical role in terms of leveraging other forms of finance. This would be where reforms to improve the effectiveness of service delivery and lowering of capital costs would be most needed.

In the context of the financial and economic crisis and constrained public budgets, however, there is a substantial risk that investments in water and sanitation services might be delayed, due to a lack of available financing. Such delays would lead to deferred benefits and potentially higher investment costs in the future, which would therefore translate in false economies.

To avoid such counter-productive reductions in funding, it would be critical to increase policy makers and funders' awareness of the substantial benefits of investing in water and sanitation services. This would also require identifying areas for priority investment, depending on where the highest benefits are likely to stem from and where the most cost-effective interventions can be identified.

Ultimately, the water and sanitation sector must include a full range of financing approaches, making the most of potential efficiency gains, adjusting targets and combining funding from both public and private sources, in order to meet its investment needs and successfully maintain and expand service. To achieve this, policy makers and water service providers need to engage in a process of strategic financial planning so as to identify what needs to be financed, how much additional resources can be generated from existing sources and how the performance of utilities can be improved to generate such efficiency gains and mobilise external financing. The set of tools presented in the next Part of this report can help in achieving such goals.

Notes

1. According to IBNET, non-revenue water represents water that has been produced and is "lost" before it reaches the customer (either through leaks, through theft, or through legal usage for which no payment is made). IWA distinguish between non-revenue water (%) and unaccounted for water, with the latter not including legal usage that is not paid for. The indicators are usually measured in m3/connection/day. The difference is usually small, and the IBNET Toolkit therefore only uses non revenue water as an indicator.

2. As noted in section 1.1, households are significant contributors of finance, particularly for non-networked systems in which they are often required to invest themselves. These financial flows are often difficult to track reliably. Where data is available, however, it would usually be included under "tariffs", which therefore captures various types of "household finance" (when users of the service pay) as opposed to public finance (when domestic and international taxpayers pay).

3. In August 2010, Manila Water was up 9.84%, Thai Tap Water Supply up 16.36% and Darco Water Technologies (Singapore) up 33.3%.

4. See p. 20 of OECD (2009c) for a description of the risks involved in WSS projects.

5. Note that households can be also considered as private actors, either when they invest in building and maintaining water and sanitation facilities on the private domain to enable them to benefit from connection to full public services or when such services are self-supplied (as with the case of on-site sanitation systems for example).

6. The contribution of the formal "private water and sanitation service operators" constitutes what is usually referred to as Private Sector Participation (PSP) in the sector.

7. Note that, in cases where households or private actors provide services without government instruction in order to fill an "access deficit" gap, they would tend to finance all necessary investments themselves either from their own funds, borrowing from family or friends, local money-lenders or microfinance institutions where they exist.

8. See *http://ppi.worldbank.org/*. It should be noted, though, that a weakness of the PPI database is that it does not provide a comprehensive picture of private investment in water infrastructure since it fails to cover some of the deals that involve mainly domestic players as well as some of the re-financing that occurs over the life-span of PSP contracts. As a result PPI numbers may fail to capture an important new source of private investment and distort the overall picture.

9. Gassner, K, Popov, A. And Pushak, N. (2009).

10. Different PSP contracts entail different impacts on operational efficiency and therefore on access to finance. This can be explained by the various risk-sharing and responsibility-sharing arrangements across different contracts, *e.g.* concessionaires are responsible for both operations and investments, while lease contracts give direct incentives to increase operation efficiency through the revenue structure.

Part II

A toolbox to support effective water and sanitation policies

Chapter 4

Introduction to the toolbox

> *To provide support to governments and water and sanitation service providers, the OECD (in conjunction with a number of other international organisations) has developed a series of tools, including financial tools, benchmarking tools and guidelines with a view to improve the performance of utilities. The audience for these tools varies, and may include policy and decision makers, municipal government staff, water utility managers, staff of international organisations, etc.*

The "Toolbox" contains a series of brief notes on each of the following tools, as briefly summarised below:

Strategic Financing Planning, based on the FEASIBLE tool. Strategic Financial Planning (SFP) is a methodology designed to help many developing and transition countries that need to engage in a reform process for the water and sanitation sector with the definition of achievable targets and financially sound planning, taking into account limited public funding. FEASIBLE is a computer-based tool that can assist with the process.

The Financial Planning Tool for Water Utilities (FPTWU) was created to assist water utilities, originally in EECCA countries, with achieving medium and long-term operational and financial sustainability through solid investment planning as well as forecasting tariffs and subsidies. It is a computerised model that allows users to summarise key technical, financial, operational parameters of a water company, calculate a set of performance indicators for utility monitoring and analyse the financial gap to meet these performance indicators on the basis of cash in and cash out. The resulting gap that needs to be filled is presented graphically and the model allows defining a program of measures in order to close the financing gap, including through tariff adjustments and/or public subsidies for capital improvements.

The Multi-Year Investment Planning Tool for Municipalities is targeted at municipalities to help them prioritise their investments in the economic and social sectors under their responsibility. To do so, the tool gathers data on historical budget trends, planned expenditures, available resources and cost of debt. Based on investment prioritisation criteria, it then sets priorities for the next 4 to 6 years.

The Guidelines for Performance-based contracts provide guidance on preparing, negotiating and implementing performance-based contracts. They include the choice of performance indicators, tariff structures and mechanisms for monitoring and enforcing the contract. This tool was developed primarily for countries in Eastern Europe, Caucasus and Central Asia (EECCA) but could potentially be applicable in other regions.

The Water Utility Performance Indicators (IBNet) is a benchmarking tool developed by the World Bank that promotes international benchmarking of water utilities and provides guidance on data collection and monitoring.

The Checklist for Public Action for Private Sector Participation in Water Infrastructure seeks to assist policy makers in assessing and managing the implication of PSP in the water sector. It identifies key policies needed for performing cooperation and provides a set of tools and practices to address those issues, based on countries experiences.

In the following chapters, each tool is presented based on a common outline:

- A first section providing some background on the type of issues and difficulties that the tool aims to address;
- A second section briefly describing the tool and summarising its main components;
- A third section giving examples of where the tool has been applied and how;
- A fourth section drawing out lessons from these applications and outlining what future applications of the tool are envisaged;
- A fifth section giving information on how to get started for using the tool and where to find more information.

Chapter 5

Strategic Financial Planning for WSS at national or regional level – the FEASIBLE tool

Strategic Financial Planning (SFP) is a methodology designed to help developing and transition countries that need to engage in a reform process for the water and sanitation sector with defining achievable targets and financially sound planning, taking into account limited public funding. FEASIBLE is a computer-based decision support tool that can assist with the process.

5.1. Background and rationale

Severe financial constraints call for a renewed approach to strategic financial planning for water supply and sanitation. The critical situation in the water supply and sanitation (WSS) sector in many developing countries and transition economies calls for a fundamental reform in the approach to financing water supply and environmental infrastructure and the associated policy, planning and institutional arrangements. Cases abound where overly ambitious plans to extend the coverage and level of WSS services need to be replaced by more realistic programmes, tailored to ensure financing for appropriate operation and maintenance, essential repairs and rehabilitation of critical elements of the WSS infrastructure, as well as sustainable extension where appropriate. Such programmes need to be cost-effective and affordable, for households and the public purse alike. Tough financial constraints are an incentive to design more efficient policies, which make the best use of available financial resources to meet realistic sector development targets, including the Millennium Development Goals (MDGs) for WSS.

Public finance requires measurable targets and monitoring progress towards achieving them. Striving to improve the cost-effectiveness of public spending and quality of budgeting, some countries have recently opted for Medium Term Expenditure Frameworks (MTEF, a 3-5-year rolling budget plus related annual budgets) and for result-oriented budgeting (where budget allocation mirrors actual delivery of policy objectives in the field). In a context where competition for public finance is fierce, these approaches call for:

- A method to demonstrate the benefits of public funding (in particular for investment) in the sector (*e.g.* contribution to poverty reduction, to social and economic development, to fiscal revenues);
- Realistic, measurable targets for the period (*e.g.* achieving the water-related MDGs by 2015);
- An instrument to assess the expenditure needs related to achieving the set targets; such an instrument should produce reliable estimates quickly and at low cost.

Strategic financial planning for WSS in developing and transition countries clearly needs to be strengthened. For instance, in the 1990s many countries in Eastern Europe, the Caucasus and Central Asia (EECCA) tried to develop target programmes for WSS infrastructure rehabilitation and development, but failed to implement them because:

- Data was lacking for robust policy analysis and policy making;

- Priorities were neither clear, nor clearly linked to policy, and investment projects were too many and too costly (unrealistic "wish lists");
- Expenditure needs much exceeded available finance;
- Policy objectives were misaligned with institutional arrangements; low management, financial and absorptive capacities created barriers for programme implementation.

Experience shows that financial planning contributes to the success of water policy reforms, especially when it is part of a broader sector planning that also includes the consideration of legal and institutional reforms. It can stimulate a policy debate on the feasibility of certain choices of policy orientation and targets (thereby helping to phase out unsustainable policies and practices), lower the costs of water policies (by appropriate targeting, sizing, and investment sequencing), and generate additional financial resources (by attracting attention of donors and financiers). The need to financially plan water supply and sanitation reforms has been better understood recently and new methods have emerged in the international arena to allocate scarce public resources across the economy.

To achieve this, a process is needed, which can be supported by a dedicated tool. To respond to this need, the OECD/EAP Task Force with financial support from Denmark has developed a methodology that comprises a process to develop Strategic Financial Plans, and a tool to process data and support the process, referred to as the "FEASIBLE" tool. A number of projects on financing water supply and sanitation, in the context of the EUWI and the EAP Task Force, build on such a process and tool developed by the OECD. They also helped to refine and further develop the SFP methodology and tool. This note captures the main features of the process and the tool that supports it.

5.2. Description of the Strategic Financial Plan process and the FEASIBLE tool

The Strategic Financial Planning process follows a three-step approach as follows:

1. **Develop a baseline scenario**
   - Carry out a robust analysis of the current situation (coverage and quality of service);
   - Evaluate initial objectives and targets (*e.g.* typically, maintaining the level of service plus implementing committed investment projects. However, if this target is politically not acceptable for the beneficiary country, a more ambitious target could be established

already for the baseline scenario, such as achieving the MDGs for water supply and sanitation);
- Define a policy package (technical, institutional, economic and financial measures) required to reach the established objectives/targets;

2. **Assess the financial feasibility of the baseline scenario**
   - Calculate the expenditure needed for implementing the baseline scenario (for implementing the measures needed to achieve the set targets, as well as operating and maintaining the system);
   - Assess available finance, taking account of the three ultimate sources of finance for water supply and sanitation (tariffs, taxes, and transfers from donors);
   - Calculate the potential cash flow gap between expenditure needs to meet the baseline scenario and available finance;
   - Assess the affordability of meeting the baseline scenario for households and other users (tariffs and user fees) and for the public budget (including capital expenditure programmes, operating subsidies and social support measures financed from public funds);
   - If a cash gap exists and/or the suggested policy package underlying the baseline scenario is not affordable, revise the policy package (*e.g.* by defining alternative targets, bringing forward cost saving measures, rescheduling investments and/or by mobilising additional finance).

3. **Develop alternative scenarios**
   - If this iterative process allows for gradually closing the baseline financing gap with an affordable policy package, then more ambitious development targets could be considered under alternative scenarios.
   - If the gap cannot be closed in any way and the present level of service is not affordable, then options for scaling down the existing level of services (sometimes referred to as "strategic disinvestment") should be considered.

The process of scenario development is an iterative process of adjusting targets, policy measures to achieve them (including deadlines), and financial means. The outcome of this iterative process can be a set of SMART targets and a realistic, feasible and affordable scenario to achieve them.

5. STRATEGIC FINANCIAL PLANNING FOR WSS AT NATIONAL OR REGIONAL LEVEL – THE FEASIBLE TOOL

A computer-based tool called FEASIBLE was developed to support the process

The tool is designed to assess a potential cash flow gap between financial needs to meet policy targets and available financial resources. The tool structures data collection and processes quantitative information on the costs of alternative water supply and sanitation policies and available sources of finance (from the 3Ts). It is designed to facilitate iterations, revisions of objectives and policy options.

FEASIBLE uses *generic cost functions*, adjusted for local conditions and prices – the functions often demonstrate substantial *"economies of scale"*. FEASIBLE aggregates figures for regional or national territories. It cannot be used to assess individual investment projects, *i.e.* for one city only.

The tool consists of several modules on urban and rural WSS, including sub modules on water supply and sanitation and a financial module. The general structure of the tool is presented in Figure 5.1.

An example of a key output of the model as used in Armenia is presented in Figure 5.2. The figure shows that higher collection rates and user fees can cover operation and maintenance costs after a given target year, which means that operating cost subsidies should be removed after that year. Financial surpluses can be used to pay back loans related to previous investment, save for

Figure 5.1. **Structure of the FEASIBLE tool**

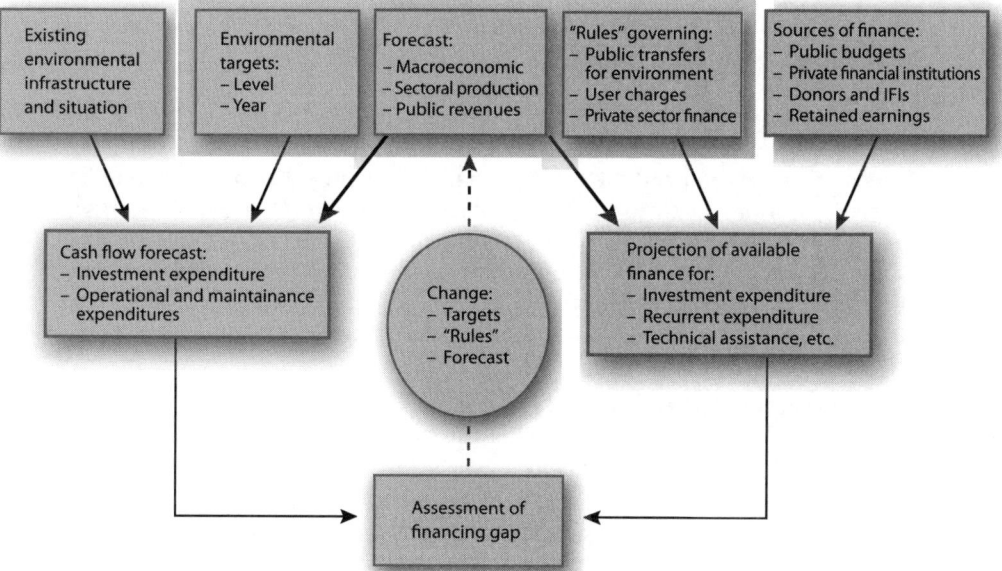

Figure 5.2. **Expenditure needs versus collected user charges in Armenia (million dram)**

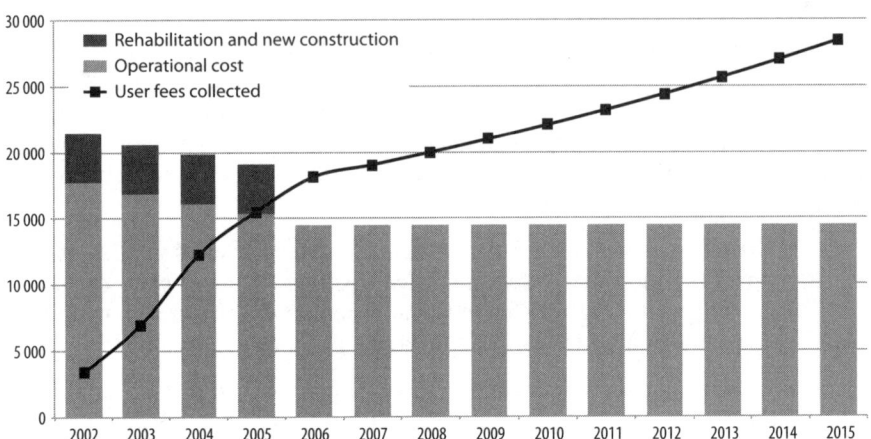

Source: OECD/EAP Task Force, Ministry of Finance and Economy of the Republic of Armenia (2004), *Financial Strategy for Urban Wastewater Collection and Treatment Infrastructure in the Republic of Armenia* (in English and Russian), prepared by COWI Moscow (*www.oecd.org/dataoecd/51/10/34596126.pdf*).

future extension, or stabilise prices. For that example, this type of analysis was used to substantiate policy decisions on investment planning and water pricing with the Ministry of Finance, operators and other stakeholders.

5.3. Where has it been applied?

The methodology was used for developing financing strategies for the WSS sector, by the OECD/EAP Task Force and DEPA/DANCEE, the European Commission and the EUWI Finance Working Group. The countries where the tool has been used include Armenia, Bulgaria, Egypt, Georgia, Kazakhstan (at the national level and in one province), Kyrgyzstan, Lesotho, Moldova, Turkey, Russia (in 6 provinces), Ukraine and some others. Lessons can be learned from these experiences on how to implement the methodology. There are plenty of opportunities to refine and expand the methodology.

5.4. Lessons learned and the way forward

The methodology essentially supports a dialogue on improving sector policy. As a result, it has to be sector policy oriented: the ultimate objective is to inform the reform process and to support decisions on such issues as coverage and quality of service, investment plans, water pricing, institutional reforms. Ideally, strategic financial planning should be part of broader sector planning that includes the consideration of legal and institutional reforms.

Second, it is an opportunity to build political support for water reforms. Application of the methodology is not a technical exercise: it involves all relevant stakeholders into a reflection process and sectoral policy dialogue. In particular, decision makers from the budgetary institutions need to be involved. Experience suggests that the policy discussion gains extra leverage when major financing donor agencies participate as well.

Third, the methodology can be used to foster implementation of policy reforms. It can translate into implementation on the ground when it is linked to decision making, *e.g.* through budgetary decision making, donors co-ordination, and dialogue with the private sector (where appropriate). The process also leads to associated policy and institutional arrangements for mobilising and allocating financial resources, and to measures to develop sufficient management, financial and absorptive capacities. Additional complementary measures experienced in case studies include adopting legislation on service quality standards, or tariff setting rules and procedures, developing institutions and building capacity.

Typical outcomes from real case applications of the methodology have included:

- *A shared understanding of issues by all key stakeholders*. For example, a national policy dialogue on Financing Rural Water Supply in Armenia helped identify realistic policy objectives for minimal water supply standards for rural populations, which are being incorporated in the legislative framework.

- *A consensus on realistic and affordable WSS infrastructure development targets*. For instance, Georgian authorities had to scale back ambitious plans regarding service coverage and service levels in medium-sized cities.

- *More objective discussion of tariff policy.* For example, in Armenia, analysis has shown that a certain level of tariffs and collection rate would allow covering operation and maintenance costs of the service in urban areas after a transition period. Additional analysis confirmed that this level of tariff was affordable for 90% of the population and helped design targeted measures to support the lowest decile that could not afford paying their water bills as well as other essential goods and services.

- *Improved dialogue with the Ministries of Finance and Economy.* For example, an ongoing project in Moldova is helping to translate a financing strategy into medium term expenditure framework, linked with budgetary decision making.

- *Inscription of results into the policy framework, including PRSP or country development strategy.*

The OECD supports the idea that more can be gained from sharing experience with the development and deployment of the process and tool, while a number of countries can benefit from the wider deployment of strategic financial planning for water supply and sanitation. A number of avenues are being explored to adapt the methodology and expand its application based on experience accumulated:

- *Continue to bring revisions to the strategic financial plans.* Plans are living documents and there are some benefits in revisiting them, to take account of new developments (demographics, new information on climate change impact on water regimes, technical innovations, economic development, restrained availability of public finance…);
- *Replicate the methodology in other countries/regions;*
- *Develop links with budgetary decision making and project finance.* The methodology can inform investment plans, which translate into budgets. In some countries, the process has been linked to budgetary decision making and the outcomes translated into multi-year framework for public expenditure, ensuring consistency between budget allocation and reform plans;
- *Adapt the process to the reform of water sector governance and water resource management, as the focus till now has been on financing water-related infrastructure.*

While the experience with strategic financial planning for water and sanitation is broadly positive, it will not lead to positive results in any type of context. Experience suggests a number of key success factors.

- SFP must be led by a champion and fully owned by host country institutions, supported by their government at a suitably high level. This applies particularly to the engagement of stakeholders both in and outside the water sector, civil society and international donors;
- The objectives of SFP have to be specific, realistic and linked to other relevant policy formation. This link implies that SFP needs time in order to engage stakeholders in a medium/long term process;
- The methodology and modelling used to develop the sector analysis must be credible and fully endorsed by all major stakeholders, including the Ministry of Finance. This implies an appropriate level of sophistication, the choice of hard data, and a continuous balancing of expenditure needs with available financing;

- For best effect SFP should be closely aligned with existing institutional arrangements for sector policy making;
- SFP should be supported actively and flexibly by donors, who should adapt their sector strategies to the outcome of the FS and be prepared to support its implementation. This will often imply support to enable beneficiaries to engage in horizontal policy dialogues (involving several authorities and civic society) and to perform policy analysis using the methodology and models of the SFP;

5.5. How to get started?

The process and the FEASIBLE tool are in the public domain and it is hoped that more experience in this area accumulates and is shared in a wider community. The FEASIBLE tool and associated User Guide can be downloaded free of charge at *www.cowi.com/feasible*. Additional technical support may be needed to adjust the FEASIBLE tool to the local context, however, as the tool in itself can be relatively complex to use.

The process can be initiated by a variety of stakeholders (usually, ministries in charge of water reforms, ministry of finance, or donors' community). The first step is to gather a large constituency of stakeholders who share an interest in water reform (although not necessarily a vision about the reform) and who commit to participate in a 1-2 year reform process. Technical assistance is required, to facilitate the process and substantiate it with data and information, and to run the tool.

The process is described in a number of documents published by the OECD and available through the EAP Task Force (see List of references).

Chapter 6

Financial planning tool for water utilities

Few water utilities have the capacity to rely on strategic financial planning. The Financial Planning Tool for Water Utilities (FPTWU) was therefore created to assist water utilities, originally in EECCA countries, with achieving medium and long-term operational and financial sustainability through thorough investment planning. It is a computerised model that allows users to summarise key technical, financial, operational parameters of a water company, calculate a set of performance indicators for utility monitoring and analyse the financial gap to meet these performance indicators on the basis of cash in and cash out. The resulting gap is presented graphically and the model allows defining a program of measures in order to close the financing gap.

6.1. Background and rationale for developing the tool

Municipal owners of municipal infrastructure and water utility operators are striving to provide appropriate solutions to their customers within their service areas. The sector is still progressing towards a modernised water sector but its development is hindered by a large number of factors. One of the main obstacles is the lack of funding sources. However, there are also a number of other issues that need to be addressed for the sector to develop. These typical issues are summarised below:

- Most municipalities and water utilities in EECCA countries do very little strategic planning within the water sector;

- The owners of communal service infrastructure (*i.e.* municipalities) are usually responsible for rehabilitation, modernisation, and development of the infrastructure. The owner has to approve any investment decisions made by the water utility – even if the water utility has the financing available;

- Municipal investment planning for infrastructure is often discretionary and there are often no clear and transparent criteria for appraising and prioritising investment projects;

- Very few water utilities have developed corporate development or strategic business plans;

- Municipal and water utility planners have little experience with multi-year investment planning; and

- Adequate regulation and tariff setting rules and procedures are often lacking, which means that tariff setting often becomes a highly politicised process (due to affordability/social concerns).

The overall idea and objective behind the development of the **Financial Planning Tool for Water Utilities (FPTWU) Tool** was to assist the water utilities in the EECCA region with achieving medium to long-term operational and financial sustainability. Such objectives, can however, only be realised when water utilities are equipped with instruments for financial modelling, planning, and analysis. Thus, the FPTWU Tool has been developed to serve as a tool for water utilities when initiating financial planning in their companies. More specific objectives of the FPTWU Tool are:

- To assist the water utilities with improving their financial status through introducing medium term financial planning, including capital improvement programming and tariff setting;

- To improve the investment decision making process through the incorporation of potential investment projects into the general framework of the financial planning process;
- To improve operational performance of the water utility through better monitoring of relevant performance indicators *e.g.* water demand, operating costs, collection rates, etc.;
- To assist with the establishment of efficient communication between water utilities and municipalities on capital improvement program/ development targets, needed tariffs and subsidies, by ensuring that communication is based on factual and well-justified information, data, and analysis through the use of the tool's outputs.

6.2. Description of the tool

The FPTWU is a computerised model (it consists of 34 interlinked Excel tables) with the support of an external database. The main modules are Data Input, Calculation Module and Output and Results Module, as shown in Figure 6.1.

Figure 6.1. **Overview of the FPTWU model**

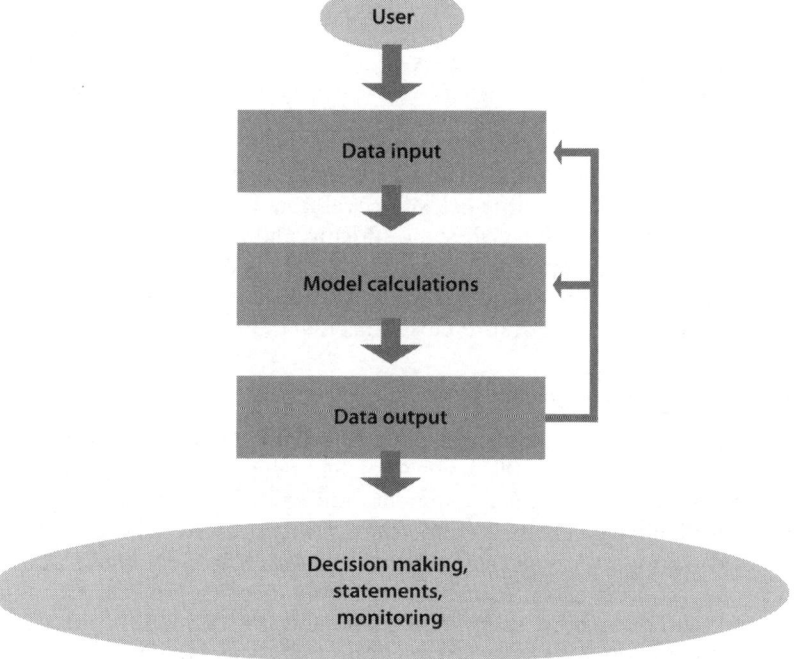

The planning period can be up to 20 years and all calculations can be carried out using either nominal or real figures. All the outputs can also be graphically displayed. Figure 6.2 represents the FPTWU model structure. It shows the inter-linkages between different modules and worksheets and establishes a relative hierarchy of the model in the form of Input-Calculation-Output.

Figure 6.2. **Architecture of the FPTWU model**

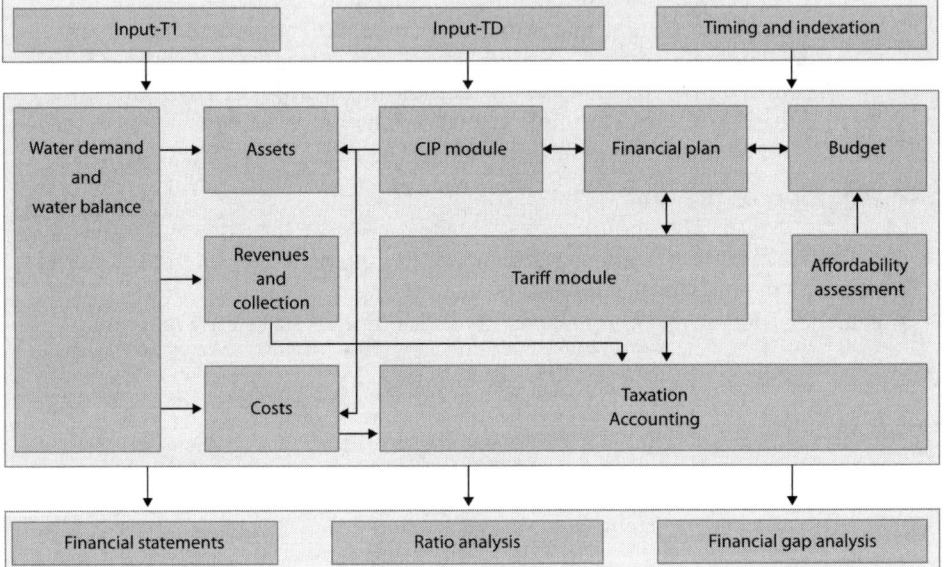

It is a computerised model that allows users to summarise key technical, financial, operational parameters of a water company, calculate a set of performance indicators for utility monitoring and analyse the financial gap to meet these performance indicators on the basis of cash in and cash out. The resulting gap is presented graphically and the model allows defining a program of measures in order to close the financing gap.

6.3. Where has it been applied?

The tool was developed and pilot tested in the Kyrgyz Republic for Bishkek water utility in 2005. The tool has been further improved based on comments received after implementation of the pilot project in Bishkek and on the feedback received from water utility experts who attended two regional dissemination workshops held in Bishkek in September 2005 and in Moscow in December 2005. Two additional pilot projects in regional water utilities (Armvodokanal) in Armenia and in Chisinau (Moldova) were completed in October 2006 and September 2007 respectively.

> **Box 6.1. Implementation of FPTWU at Bishkek Water Company (the Kyrgyz Republic)**
>
> The first practical implementation of the FPTWU Tool took place at Bishkek Water Company. The underlying objective was to develop a company-tailored financial management instrument which would assist Bishkek Water Utility in implementing sound medium to long term financial planning, provide a systematic and complete overview of the financial and economic situation of the company, improve the investment decision making process, and assist in preparing well justified investment proposals to investors and lenders.
>
> The work was carried out mostly at the premises of Water Company with close involvement of various staff members from technical, financial, economic, operational and other departments. A substantial amount of data and information was processed and the draft FPTWU model was adopted and fine-tuned to correctly reflect the situation at the utility. On the basis of the approved tool, as well as on the basis of the lessons learned during the implementation process, a training manual was prepared and a training workshop was carried out with participation of not only Bishkek Water Company staff, but also representatives of other water utilities in the Kyrgyz Republic and other EECCA countries.
>
> The FPTWU Tool and the entire process of its development had an important impact on Bishkek Water Company. It helped to enhance financial management practices of the water company and orient it towards a medium to long term planning horizon. During the implementation process a large amount of previously scattered company data was accumulated in one place, allowing a more integrated and systematic overview of the situation of the Company. With the use of the FPTWU Tool it became easier to formulate and analyze the financial sustainability of various investment and infrastructure rehabilitation projects. Bishkek Water Company used the tool to develop a rehabilitation and investment plan which was later on financed by EBRD and is currently under implementation. Finally, the process of implementing the FPTWU Tool helped to substantially improve the capacity of the financial and economic management staff of the water utility, enhancing their ability of utilising modern approaches and concepts of financial analysis and planning.

6.4. Lessons learned and the way forward

The use of the FPTWU model for these water utilities resulted in:

- An integrated medium to long term financial plan covering all financial aspects of water utilities;
- The definition of a medium to long term financial plan including a capital improvement program, design of tariff policy and identification of the need for public subsidies, based on the objective to achieve financial sustainability for the water utility;

- The enhancement of the water utility's credibility in relations with the clients, owners and financial institutions.

The primary target group for use of the FPTWU Tool is the financial planning department or financial planning specialists in water utilities. The Tool and its functions/properties can also be utilised by the economic, financial departments of municipalities. In particular, it can be used for revision and approval of water and wastewater tariffs as well as for allocating municipal budget funds to water utilities. In such cases, however, the municipal department's role will be that of a "user" of the Tool. In other words, while water utility specialists can develop the tool's output, through filling-in of the necessary input data, municipal specialists can utilise the tool's output in order to provide a qualitative basis for their decision-making.

6.5. How to get started

Additional information about the tool can be found at the OECD web-page: *www.oecd.org/document/15/0,3343,en_2649_34343_42958607_1_1_1_1,00.html*

Chapter 7

Multi-year investment planning tool for municipalities

Long-term capital planning is required to expand and repair water and sanitation infrastructure. The MYIP tool is targeted at municipalities to help them prioritise their investments in the economic and social sectors under their responsibility. To do so, the tool gathers data on historical budget trends, planned expenditures, available resources and cost of debt. Based on investment prioritisation criteria, it then sets priorities for the next 4 to 6 years.

7.1. Background and rationale for developing the tool

The ability to undertake long-term capital improvement planning directed at meeting the infrastructure needs of the community is of crucial importance for the future economic and social development of communities. The project preparation stage itself takes longer than one year for most infrastructure projects and when the implementation period is also considered, infrastructure investment planning has to be long-term. A tool that helps local governments to develop a long-term financial vision is the Multi-Year Financial Plan.

The multi-year investment planning tool (MYIP) is a rational, rule-based approach to investment planning by local governments, supported by a financial planning software. Rather than looking into accounting systems, the MYIP works with already existing investment plans combining and prioritising them. The program is a computer application designed to facilitate multi-year investment planning in municipalities in many aspects: it enables the creation of a multi-year financial plan through an analysis of historic budget trends and setting growth coefficients for each budget element, creating and prioritising a list of investment tasks as well as a computer-assisted simulation of various options depending on prioritisation parameters, financing methods and implementation dates.

Using the MYIP tool can help municipalities with achieving the following:

- Introduce mid-term capital improvement planning;
- Adopt a multi-year planning perspective as opposed to one-year planning;
- Select capital projects on an objective and transparent basis;
- Develop an investment strategy;
- Communicate with citizens about the most important strategic investments;
- Communicate with banks and financing institutions to apply for funding.

This planning tool is based on the Capital Improvement Planning (CIP) approach used in OECD countries (for example, in the United States), not only by local governments, but also by many other institutions.

7.2. Description of the tool

The multi-year investment planning tool provides a process for selecting strategic investment projects in a long-term perspective that achieves the largest possible benefits (financial, social, ecological and others) as a result of their implementation. The Multi-Year Financial Plan includes the following elements:

- Local government revenue and operating expenditure forecast;
- Service and repayment of liabilities incurred;
- Amount of debt to be incurred;
- Total funds intended for investments projects;
- Clear selection criteria and investment project priorities;
- List of investment projects to be implemented, their scope and sources of financing year by year.

Using the tool can help the local government with identifying the investments that it should implement in the next four to six years, by identifying which investments are beneficial to local society, which are less important, which can be delayed for the future and which are bad ideas that should be abandoned. It then helps the government assess whether it will have sufficient money for financing the plan each year, whether it should take out a loan for this purpose and whether it will be able to repay it. Capital Investment Planning include water and sanitation services among other municipal services.

The preparation of the Multi-Year Financial Plan (MIP) is done through the following successive stages:

- **Stage I: analysis of past data.** The goal of this stage is to analyse historic budget data as well as to evaluate the city in terms of its financial and investment potential. The analysis focuses on budget lines that have been stable in the past, because they provide the basis for future projections. This allows forecasting revenues and expenditures for the planned years.

- **Stage II: forecast of city revenues and expenditures, as well as investment potential over the planned period of analysis.**

- **Stage III: analysis of city current debt and ability to incur further debt.** The goal of this stage is to define the city's capacity to incur debt.

- **Stage IV: presentation of the city's planned investments.** The goal of this stage is to describe the characteristics of all investments that the city is planning to implement in the analysed period.

- **Stage V: determination of rules for selecting investments for implementation.** The goal of this stage is to define a list of investments ranked from the best to the worst according to the accepted procedures for classifying and prioritising investments.

Figure 7.1 shows the various steps and how they are linked together. It also shows the inputs and outputs of the various steps. The final output is the "Final Document", which is a Multi-year Investment Plan approved by the city council.

7.3. Where has it been applied?

Multi-year Investment Planning has been successfully implemented in a number of local governments in the Central and Eastern European countries, *e.g.* in Russia and Ukraine (see Box 7.1). Given the challenges that Eastern European local governments face, MYIP was selected as a tool that could help improve investment planning. Low and unpredictable revenues, rent-seeking in investment planning and inter-budgetary transfers and limited access to loans and credits are just a few of the problems that can be alleviated through better financial planning and more transparent project selection.

7.4. Lessons learned and the way forward

Where it has been applied, the MYIP tool has allowed achieving the following results:

- Decreased influence of the political cycle on the capital improvement process in the city, in particular the capital project selection process;
- Improved planning and faster implementation of capital investment projects;
- Reduced practice of formulating overstated needs by departments/ municipal companies;
- Tool for communication with the local community;
- Facilitated access to external sources of finance that require a MYIP (almost all significant ones require some form of investment plan) and an evaluation of creditworthiness;
- Identification of potential for co-financing by the municipality itself.

7.5. How to get started

Additional information about the tool as well as case studies can be found at the OECD web-page: *www.oecd.org/document/47/0,3343,en_2649_34335_ 35193199_1_1_1_1,00.html#Training_Materials_for_MYIP*

Figure 7.1. **Steps of the Multi-Year Investment Planning Process**

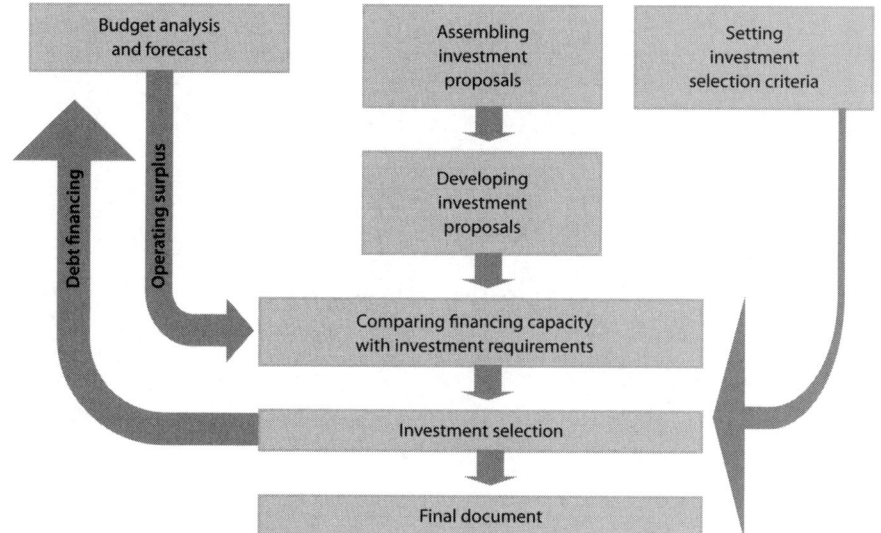

Box 7.1. **MYIP implementation for the City Lutsk in Ukraine**

Successive stages of developing the Multi-Year Financial Plan for the City of Lutsk brought into sharp relief the large investment needs associated with municipal infrastructure on the one hand and the limited potential for the city to finance these investments by itself on the other.

In the City of Lutsk, the Multi-Year Financial Planning exercise reflected a certain vision for the future of the city by local government officials. The top ten investments ranked according to the criteria of importance were investments associated both with improvement of municipal infrastructure (heating systems, water mains) as well as cultural and historic infrastructure (*e.g.* rehabilitation of the old town). Upon analysis of the remaining investments, it was possible to assert that these types of actions would be favoured in years to come. The city authorities accepted in full both the procedure and format in which work on the investment plan was conducted, as well as the results of these efforts. The Executive Committee of the Lutsk City Council later passed a resolution approving the Multi-Year Financial Plan.

Chapter 8

Guidelines for performance-based contracts

In many developing and transition countries, the performance of water utilities is poor and many countries have therefore turned to performance contracting. The OECD has developed Guidelines in which the key elements for preparing, negotiating and implementing performance-based contracts are addressed. They include the choice of performance indicators, tariff structures and mechanisms for monitoring and enforcing the contract. This tool was developed primarily for countries in Eastern Europe, Caucasus and Central Asia (EECCA) but could potentially be applicable in other regions.

8.1. Background and rationale

Since the break-up of the Soviet Union, the countries of Eastern Europe, Caucasus and Central Asia (EECCA) have undertaken significant economic and market reforms, including in the water supply and sanitation sector. Despite the reforms, however, the governments of these countries are still experiencing serious challenges in providing high-quality water services to their population. The poor state of water infrastructure in EECCA is a result of many years of neglect and under-investment as well as inefficient management practices.

To improve the performance of their water utilities, many countries have turned to performance contracting. If designed properly, performance-based contracts between municipalities and water utilities can help lay the basis for the long-term sustainability of water utilities, increasing their efficiency and creating conditions where investment capital can be attracted. Generally, performance-based contracts are developed to help define the utility development goals and include time-bound performance targets against which the performance of the operator is measured.

To support EECCA authorities that are willing to contractualise their relationship with their water utilities, the OECD developed "Guidelines for Performance-Based Contracts between Municipalities and Water Utilities in EECCA". These Guidelines address the key elements that need to be considered in connection with the preparation, negotiation, implementation, and periodic revision of a successful performance-based contracting mechanism.

The good practices identified in the Guidelines have been subsequently tested in a number of EECCA countries. In practice, this implies analysing existing contracts (or contracts under preparation) against the international benchmarks contained in the Guidelines and developing recommendations for the contracts' possible improvements. The case studies, carried out in Armenia, Kazakhstan and Ukraine, broadly cover the main types of performance-based contracts that exist in the water sector, including, management, lease and concession contracts as well as a case of full divestiture. The lessons learnt from these case studies were then summarised and an updated version of the Guidelines was prepared.

8.2. Description of the tool

The Guidelines provide a general framework for designing performance-based contracts in the water sector and identifies good international practices and standards for such contractual arrangements. The Guidelines address all major elements that need to be in place when national or local authorities design performance-based contracts. The main issues covered in the guidelines include among others:

Contract preparation

- Choice of contract type and contract duration
- Review of the legal and regulatory framework
- Review of the utility's assets and liabilities – restructuring of the utility
- Preparation of the bidding and selection process
- Accuracy of initial data and information

Performance indicators

- Definition and selection of indicators
- Definition of the baseline scenario
- Monitoring of performance indicators
- Choice of a technical auditor

Tariffs and financial obligations of the contracting authority

- Types of tariff structures
- Mechanisms for tariff adjustment

Financial obligations of the contracting authority

Monitoring of contract implementation

Mechanisms for conflict resolution and contract enforcement

Risk management

- Types of risks
- Risk allocation principles

Personnel management

To analyse individual contracts, an evaluation methodology, based on the recommendations provided in the Guidelines, was developed. The methodology consisted of a detailed questionnaire coupled with direct interviews with relevant stakeholders in the countries. The findings and recommendations emerging from the reviews were then discussed and agreed upon at stakeholder meetings.

It should be noted that the Guidelines do not aim to deliver a complete, "ready-to-use" toolkit for immediate application. The good practices and approaches proposed in the Guidelines need to be further adjusted and tailored to the needs of the individual municipality and type of contract. Which of these approaches will be used by a given municipality will depend on the governance structure in the country as well as the maturity of the parties involved. In addition, it is also important to note that the Guidelines do not aim at substituting legal advice on regular contracting: they rather seek to provide additional expertise based on international experience with performance contracts in the water sector.

8.3. Where has it been applied?

The good practices identified in the Guidelines have been tested in Armenia, Kazakhstan and Ukraine. The case studies include a management contract (for the Armenia Water and Wastewater Company with the French company SAUR) and a lease contract (for the Yerevan Water Supply Company with the French company Veolia Water) in Armenia. In both cases in Armenia, the testing was done after the contracts were signed, as described in Boxes 8.1 and 8.2. In addition, testing of the Guidelines was carried out in the context

---

Box 8.1. Lease contract for Yerevan Djur, Armenia

In December 2005, the Armenian government awarded a lease contract to the French company Veolia Water to manage the water utility in the city of Yerevan. The contract was awarded for a period of 10 years. To implement the contract, the bidder created a new company, Yerevan Djur, which is wholly owned by Veolia Water. Under the lease contract, the operator pays the lessor a fee on a semi-annual basis for the period of the contract.

Under the lease, the private operator is responsible for operating and maintaining the utility and more specifically for providing water and wastewater services to the population of Yerevan municipality. There are wholesale and retail tariffs for water supply and wastewater. The tariff level was agreed upon between the government and the operator at the start of the contract over the period of contract duration. There is also a possibility to adjust the tariff on an annual basis taking into account such parameters as inflation, exchange rate fluctuations, changes in the electricity tariff and in the level of water consumption. The tariff is set to cover all operation and maintenance costs, excluding investment and depreciation costs. The lease contract includes a number of performance indicators as well as penalties if the operator does not meet the indicators.

The government, on the other hand, is responsible for financing investments. To ensure investments for the Yerevan water utility, the government contracted a USD 18.5 million loan from the World Bank. The revenue from the lease fee paid by the operator is used to pay back this loan. However, it is the operator's responsibility to plan, design, tender and supervise works financed with resources of the World Bank. Any new assets built with the World Bank resources remain a state property but they are handed over to the operator to manage during the period of the contract.

of two concession contracts in Ukraine (with domestic private operators in the towns of Berdyansk and Kupyansk) and a divestiture in Kazakhstan (the water utility in the city of Shymkent is owned by a domestic private operator).

In both cases in Armenia, the testing of the Guideline was done after the contracts were signed, as described in Boxes 8.1 and 8.2. The reports from the reviews of the two contracts were discussed at stakeholders' meetings, including representatives of the government, the private operators, the NGO community and consumer protection organisations. This allowed all the parties to better understand the premises of the contracts as well as, for example, reach consensus, on the basis of international experience presented in the analysis, on certain methodological issues related to measuring individual performance indicators (*e.g.* the water quality indicator).

While the two Armenian contracts were generally well-designed and would meet international standards for such contracts, the authorities in

---

Box 8.2. Management contract for the Armenia Water and Wastewater Company, Armenia

The management contract for the Armenia Water Supply Company Service Area was signed in August 2004 between the Armenia Water and Wastewater Company (AWWC) and the French company SAUR SA for a period of four years. The contract was additionally extended in 2008. The contractor is paid a fixed fee, on a monthly basis, out of a World Bank loan. The contract envisages a number of performance indicators for the operator to meet. While there are no penalties foreseen, the contractor can receive an additional bonus payment depending on the level of achieved performance.

SAUR provides services to 10 regions in the country (37 towns and 280 villages or about 700 000 people). The operator has full responsibility for the management, operations and maintenance of the water and wastewater system in the service area. All the operator's costs are financed through the tariff and government subsidies (operational deficit and investments). The tariff, approved by the Public Services Regulatory Commission, is volume-based and is identical for all users. It is split into 3 parts: tariff for water supply, for wastewater collection and for wastewater treatment. The tariff has been significantly increased since the start of the contract on several occasions.

As the World Bank initiated the project, it funded the project preparation phase and finances the management contractor's fixed fee, the performance incentive compensation and the Contract Monitoring Unit. It also finances the procurement of goods, services and works needed for the company's operations as well as investments in the networks and facilities. SAUR is responsible for designing the works to be implemented with the World Bank funds and managing the related procurement process. Given the World Bank's involvement, the basic investment strategy has to be co-ordinated with and approved by the Bank.

Kazakhstan and Ukraine particularly benefited from the application of the guidelines. When the guidelines were introduced in the town of Kupyansk (Ukraine), the concession contract was under preparation. The discussions with local decision-makers on the draft contract helped them understand the need for regulating certain issues through the contract. As a result, some elements that were originally missing or even not required by the national legislation were added to the contract (*e.g.* performance indicators, rules and procedures for monitoring of contract implementation). In the case of the Shymkent water utility (Kazakhstan), where the utility was privatised, the analysis based on the Guidelines recommendations helped identify certain improvements that could be introduced despite the lack of a clear cut performance contract. As a result, the water utility management and the local government worked together to develop performance indicators for the utility which could be used to more objectively monitor its performance as well as better justify the need for tariff revisions.

8.4. Lessons learned and the way forward

The major objective of these reviews was to conduct an independent and objective evaluation of all important aspects of the contracts against the criteria and benchmarks identified in the Guidelines and provide practical recommendations for the improvement of these contracts in line with international good practices. The analysis from these reviews was then used as a basis for reaching consensus among major stakeholders with regard to needed improvements in the contracts.

In addition, through the individual case studies, the reviews also sought to deepen the analysis that is contained in the Guidelines in order to further improve their relevance to EECCA municipalities and water utilities so as to ensure a wider dissemination of the best practices in this area. The preparation of the updated version of the Guidelines (including through discussions at a national, regional and international level) has helped increase the understanding of key decision-makers in the EECCA region of the potential benefits of using performance-based contracts in the water sector.

8.5. How to get started

The tool is first and foremost targeted at decision-makers with responsibilities for water infrastructure at a municipal level in the EECCA countries that are considering introducing performance-based contracts for their water utilities. Decision-makers and politicians at a national level responsible for setting water sector standards and supervising the performance of water sector operators may also be interested to learn from the experience of other countries with such contracts.

Although the main audience is decision-makers from EECCA, the major principles and approaches to designing and implementing performance-based contracts in the water sector are similar and may be relevant for countries from other regions of the world that are contemplating entering into such contracts or striving to strengthen and improve existing contracts in line with good international practices.

In addition, managers of technical assistance programmes from different donor agencies, international financing institutions (IFIs), international organisations concerned with the practical implementation of good practices in this area and consultants working on contractual arrangements in the water sector may also find the report useful in their professional work.

Additional information about the tool as well as the individual case studies can be found at the OECD web-page: *www.oecd.org/document/10/0, 3343,en_2649_34343_40689034_1_1_1_1,00.html*

Chapter 9

Water Utility Performance Indicators (IBNET)

> *One way to increase the volume of investment towards sustainable water and sanitation infrastructures is to produce clear and accurate data that allows assessing and comparing performance across utilities. IBNET is a benchmarking tool developed by the World Bank and supported by a number of donors (including DFID and WSP) that promotes international benchmarking of water utilities and provides guidance on data collection and monitoring.*

9.1. Background and rationale

The dilapidated state of urban water utilities in some transition and developing countries represents a growing public health concern and a threat to the environment (pollution of water resources and soils). This situation requires an increased mobilisation of financial resources, which will be possible only with the progress of reforms and the support of the population. One step towards a reversal of this situation is producing clear and accurate data on the conditions of the Water Supply and Sanitation (WSS) sector in general and of utility companies in particular.

In the EECCA countries, this need for general background information has been addressed by undertaking surveys of water utility performance since 2001. This raised the question of how such benchmarking data could be used in a more advanced manner so as to help with increasing the volume of investment in the WSS sector by providing:

- A standardised framework for public information (tariff increases being often a condition to the sustainability of an investment project, objective information on service quality improvements may help strengthening citizens' support);
- A management tool to identify and monitor the key areas for operational improvement (*e.g.* scope for efficiency savings). Such operational savings plans are often a condition for utilities to be able to reimburse their loans;
- A tool to identify priority areas for investment in a given utility (project design level);
- A basis for the establishment of contractual relationships between utilities and responsible bodies (usually municipalities);
- A tool for utilities to share experience and disseminate best practices among themselves.

The purpose of benchmarking is to search for and identify best practice in the water sector with the objective of implementing appropriate best practice and improving performance. The mere collection of data is not benchmarking although it is an integral step in the path towards improved performance.

IBNET provides a means and a set of tools for water and sanitation utilities to develop national or regional groupings for the purpose of undertaking regular benchmarking activities. The methodology for estimating the indicators was developed by the World Bank. IBNET provides the opportunity for these local benchmarking initiatives to undertake international comparisons by making available easy to use search and query features.

IBNET supports and promotes good benchmarking practice among water and sanitation services worldwide by:

- Providing guidance on indicators, definitions and methods for collecting data;
- Providing guidance on setting up national or regional benchmarking schemes;
- Enabling utilities to undertake peer group performance comparisons;
- Facilitating access to water utility performance data in the public domain.

Benchmarking at a local, national and international level helps water and sanitation utilities to find comparators for identifying and sharing best practice, new knowledge and in ensuring that nothing is missed in the important job of delivering water and sanitation services to their customers. IBNET has an important role to play as a facilitator for the sharing of best practice between water and sanitation utilities across the world and for providing the information for all those working in the sector, be they funding agencies, consultants, academics and most importantly the water and sanitation utility managers.

9.2. Description of the tool

The IBNET Toolkit (Figure 9.1) has been developed to support the above concept and to provide initial support to newly establishing benchmarking schemes. The IBNET Toolkit is available in English, Spanish and Russian and includes the following resources:

- A set of core indicators (26) on which stakeholders can build their own customised measurement and monitoring system;
- A data list complete with robust data definitions;
- A data capture system that also calculates the complete indicator set;
- A method to share information on benchmarking.

Figure 9.1. **Overview of IBNET data sets**

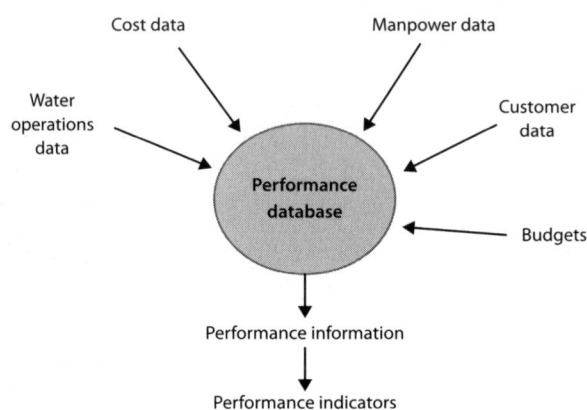

Source: www.ib-net.org.

9.3. Where has it been applied?

The overview of current benchmarking practices in EECCA countries shows that while benchmarking data are rarely used for making investment decisions and setting performance targets, the practice of performance data collection is widespread. The main problem lies in the fact that such data is available in most EECCA countries but is collected and kept most often for its own sake. Actors of the WSS market possess a potentially powerful instrument but not the skills to employ it.

Demand for performance indicators by the actors in WSS sector is therefore limited as of today, but likely to increase significantly in the near future. The need for more efficient regulation and investment monitoring will stimulate the demand for performance indicators from local authorities and regulators. Another important driving force will probably be the growing private sector participation in the provision of WSS services. Private companies are interested in a greater transparency of the WSS sector and believe that benchmarking data will discriminate the innovative players in this market.

9.4. Lessons learned and the way forward

The provision of comparative information and its use in benchmarking becomes an important management tool for managers and professionals in water and sanitation utilities. Benchmarking and knowledge of best practice is important for all water and sanitation utilities:

- Benchmarking helps managers to understand the performance of their utility relative to others;
- Benchmarking facilitates the sharing of best practice information and supports decisions to improve performance.

If data definitions are shared and used by a sufficient number of participants, at least for a core set of indicators, this network will add value to all its users and contributors by providing them with useful international comparative information.

9.5. How to get started

Additional information about the tool can be found at the web-page: *www.ib-net.org*

Chapter 10

Private sector participation in water infrastructure – checklist for public action

> *Many governments have turned to private sector participation to inject capital or improve the performance of utilities. However, the implications of involving the private sector may be sources of misunderstanding as highlighted in section 3.4. The* Checklist *precisely seeks to assist policy makers in assessing and managing the implication of PSP in the water sector. It identifies key policies needed for performing cooperation and provides a set of tools and practices to address those issues, based on country experiences.*

10.1. Background and rationale for developing the tool

Over the last 20 years, many governments have turned towards the private sector to inject much-needed investment capital and improve the often poor operational performance of publicly-run utilities. These experiences have underlined the importance of better risk management and of strengthening the creditworthiness of the water sector.

Responding to these challenges, the OECD, working with non-OECD countries and stakeholders, has developed a *Checklist for Public Action* in the framework of its first Horizontal Water Programme. This checklist provides governments with a coherent set of policy directions that address the allocation of roles, risks and responsibilities, as well as the framework conditions necessary to make the best use of private sector participation in the sector. The guidance was developed based on the OECD *Principles for Private Sector Participation in Infrastructure* and regional consultations in Zambia (November 2007), the Philippines (March 2008) and Mexico (September 2008). It was completed in time to contribute to the 5th World Water Forum held in Istanbul in March 2009.[1]

The second phase of the OECD Horizontal Water Programme (from 2009) has sought to operationalise the OECD *Checklist for Public Action*, by helping a number of governments to implement and monitor the policy recommendations contained within it. This was done through assessments of countries' conditions for private sector participation against the *Checklist* and national policy dialogues in a number of countries including Russia, Egypt and Lebanon.

10.2. Description of the tool

The *Checklist* is intended to help governments and other stakeholders to properly assess and manage the implications of private sector participation in the financing, development and management of water and sanitation infrastructure. It does not provide a detailed approach of the steps necessary to engineer a specific partnership, but aims to offer governments a clearer picture of the multiple policy areas in which decisions have to be made when private sector participation is considered. In this perspective, the *Checklist* defines the main specificities of the water and sanitation sector that bear on the co-operation between the public and the private sector; identifies key policy issues for consideration by governments; and provides a set of available tools and practices, building on recent country experiences.

The *Checklist* highlights 24 principles, organised in five key policy areas:

Deciding on the nature and modalities of private sector participation. The *Checklist* highlights the need for governments to clarify the ultimate objectives for service provision and the contributions that the private sector can make. This involves clarifying the roles and responsibilities of the private partners and defining the modalities of their involvement so that the partnership is tailor-made to local specificities, provides the incentives for sustainable cooperation in the public interest (including through fiscal discipline and transparency) and brings value for money.

Providing a sound institutional and regulatory environment for infrastructure investment. Private sector participation does not relieve governments of their responsibility to ensure safe and efficient water services and to prevent abuses of monopoly position. This involves developing a conducive framework based on high quality regulation, competition and political commitment (including to fight corruption).

Ensuring public and institutional support for the project and choice of financing. Beyond the development of a proper legislative and regulatory framework, the *Checklist* highlights the importance of ensuring the effective implementation of regulations and contractual provisions. This involves a clear allocation of roles across responsible authorities, empowerment of these authorities through clearly defined and communicated objectives and strategies, capacity building and co-ordination mechanisms.

Making the co-operation between the public and private sectors work in the public interest. Mechanisms that improve accountability of the different stakeholders and reduce uncertainty for private actors are important in the context of long term partnerships where the contracts cannot be exhaustive. These include the strengthening of competition in the bidding process and in the contractual phase; allocating risk to the party best able to manage it; using performance-based contracts (see chapter 8); inserting clauses and mechanisms to frame the discussions on unforeseen or emerging issues as well as formal dispute resolution mechanisms; and establishing monitoring processes, based on appropriate information, combined with penalties and rewards.

Encouraging responsible business conduct. Private actors have an important role to play and responsibilities in ensuring the sustainability of partnerships and that their contributions make a difference in improving the lives of millions of people. This involves participating in good faith and with commitment, promoting integrity, communicating with consumers, and effectively managing the social and environmental consequences of their actions.

10.3. Where has it been applied?

In 2009 and 2010, the OECD has worked with Egypt, Russia and Lebanon to conduct assessments of framework conditions for private sector participation in water infrastructure against the *Checklist for Public Action* and support national policy dialogues in partnership with other processes and organisations. In early 2011, work was initiated with Tunisia and Mexico.

In *Egypt*, the assessment was carried out in partnership with the Mediterranean component of the Global Water Partnership (GWP-Med) as part of the Policy Dialogue on Water of the MED-EU Water Initiative. It led to two field missions and a meeting hosted by the Holding Company for Water and Wastewater in January 2010. Similar joint GWP-Med/OECD activity was officially launched in *Tunisia* on 23 May 2011 by the Ministry of Agriculture and Environment.[2]

In *Russia*, the assessment was developed in partnership with the Task Force for the Implementation of the Environmental Action Programme, Russian institutions (Vnesheconombank and the Ministry of Economic Development) and international partners (EBRD and the World Bank). It led to two dialogues held in-country, including the Regional Meeting on Private Sector Participation in Water Supply and Sanitation in EECCA (January 2010) and the Policy Dialogue Meeting on Water and Sanitation Policy in Russia (June 2010).[3]

In *Lebanon*, the assessment was developed in the framework of the Med-EUWI policy dialogue on water, together with GWP-Med and the Ministry for Water and Energy. The activity was launched in March 2010 by the Director for Water at the Workshop on Private Sector Participation in Water Infrastructure in Lebanon organised by the Ministry, led to a first discussion at the occasion of the October 2010 Lebanon Water Week and to a national workshop on private sector participation and the role of private banks in December 2010.

In *Mexico*, joint Conagua/OECD activity using the *Checklist* was officially launched at the Water Dialogues held in the margin of the UNFCCC COP16 in December 2010 and led to a first workshop on Conagua premises in February 2011. The assessment will be carried out in 2011 as part of a broader policy dialogue in support of the development and implementation of the Water Agenda 2030.

All assessments follow a similar structure and comparable preparation processes: they provide an overview of recent developments with private sector participation in the water sector, highlight areas for consideration by government and suggest ways forward. They build on answers to a questionnaire based on the OECD *Checklist for Public Action*, publicly available material, interviews with various stakeholders during OECD field missions and national policy dialogues.

10.4. Lessons learned and the way forward

A number of lessons can be derived from the use of the *Checklist* as a basis for assessment and policy dialogue, both with respect to the tool itself and to the trends in private sector participation in the selected countries.

The need for Checklist-based types of assessment is strong among developing and emerging countries. This is prompted by various reasons, including realisation of the important infrastructure gaps that such countries face, both because of underdevelopment of new infrastructure and lack of maintenance of existing ones. In Egypt, the government is committed to mobilising private sector financing and expertise for the development of wastewater treatment facilities to address raising pollution concerns and a persistent mismatch between water and sanitation coverage (the government estimates at 5.5% to 7% of its yearly GDP, some USD 13 billion, the total amount needed to cover its infrastructure needs, including water). In Russia, a rundown water network is driving the authorities to call on the private sector to bring in the necessary investments and technical and managerial capacities to improve efficiency and quality of water service provision (an estimated USD 459 billion is needed to carry out necessary upgrades, refurbishment and new construction for water and sanitation infrastructure by 2020).

The Checklist fulfils a strong demand to share tools and access good practices used elsewhere. Demand to use the *Checklist* comes from countries with diverse needs and at different – but usually early – stages of private sector participation. In Lebanon, such experience is quasi null. Interest in the *Checklist* was prompted by a wish to better understand the benefits and risks in involving the private sector. The *Checklist*-based work was a unique opportunity to start a dialogue between private banks and the government on investors' needs and risk perception. In Egypt, the first major BOT (Built-Operate-Transfer) deal in the wastewater sector was signed in June 2009 for a wastewater treatment plant in New Cairo. The New Cairo deal is seen as a pilot project, which will demonstrate the capacity of the Egyptian government to accommodate this new form of infrastructure financing. The *Checklist*-based assessment allowed a discussion on critical conditions for a sustainable scaling up of PPP projects in the country. In Russia, active penetration of private business into Russia's water sector dates back to 2003 but has recently stalled in response to diverse factors, including recent developments in the legislative framework. Interest in the *Checklist* was prompted by the need to review and solve the bottlenecks to beneficial private sector involvement. In all cases, the *Checklist* is viewed as a flexible and neutral tool, able to bring together stakeholders and support policy dialogue on a difficult topic.

In the current context, the focus of the Checklist on framework conditions is of particular relevance. Demand comes from countries where the legislative framework is under development and allocation of responsibilities across authorities is pending. In Lebanon, the legislative framework is not yet in place and current legislation does not allow private sector participation in the water sector. In Russia, after initial years where private participation in the water sector developed on an ad-hoc basis, the legislative framework is under amendment, including the concession law. In Egypt, the Law for Regulation of Public Private Partnership was ratified by the People's Assembly in June 2010. In the current context of credit constraint and tighter financial conditions, private developers are likely to be more selective, demanding higher quality, more "bankable" projects, with clearer forms of public support and risk-sharing arrangements. Inadequate framework conditions constitute a risk that private sector will factor in by attributing a cost, prompting countries to pay increased attention to framework conditions for private sector participation.

The Checklist usefully complements other existing OECD tools. In most countries under review, the development of accountability mechanisms is limited and important uncertainty remains with regards to the reliability of available information on the status of the infrastructure and the performance of providers. The establishment of performance-based contracts or competitive bidding is proving a complex process with important and broader ramifications (for instance the impact of unclear property rights in Russia). The development of a high-quality institutional framework is constrained by the lack of capacity and of co-ordination mechanisms at sub-national level. Limited financial sustainability is also proving a bottleneck shared by all countries, with low levels of cost-recovery and/or inadequate price regulation impeding the involvement of private partners. Reciprocally, PPP projects raise issues of long-term affordability for governments when they generate contingent liabilities and lock in countries in long-term subsidies. The *Checklist* highlights these areas of major importance for governments and provides a point of entry for using companion tools that can help countries overcome the challenges, such as the *Strategic Financial Planning* (chapter 5), *the Guidelines for performance based contracts* (chapter 8) and the forthcoming publication on *Water Governance in OECD Countries: a Multi-Level Approach* (OECD, 2011).

10.5. How to get started

The *Checklist for Public Action* is primarily addressed to governments and other tiers of the public sector that are responsible for the provision of drinking water and sanitation services. The *Checklist* may also be of use to other constituencies, such as the private sector, civil society (NGOs, communities, users) and the international donor community, for a better understanding of the issues at stake and as a platform for policy dialogue.

The tool as well as individual country assessments are available at *www.oecd.org/daf/investment/water.*

Notes

1. *www.oecd.org/water* and *www.oecd.org/daf/investment/water*.
2. See *www.oecd.org/dataoecd/18/62/47360935.pdf* for a detailed discussion of the Egypt policy dialogue.
3. See *www.oecd.org/dataoecd/18/59/47360976.pdf* for a detailed discussion of the Russia policy dialogue.

References

AESN (2007), "Bénéfices de l'assainissement", Rapport d'étude AESN, February 2007, Seine Normandie, France. Banerjee, S. and E. Morella (2011), *Africa's Water and Sanitation Infrastructure, Access, Affordability and Alternatives,* The World Bank, Washington, DC, April 2011.

Cairncross, S. and V. Valdmanis (2006), "Water Supply, Sanitation and Hygiene Promotion", Chapter 41 in *Disease Control Priorities in Developing Countries*, Second Edition.

DANCEE, OECD/EAP Task Force (2000), Moldova, *Background Analyses for the Environmental Financing Strategy* (in English and Russian) prepared by COWI A/S.

DANCEE, OECD/EAP Task Force (2000), *Municipal Water and Wastewater Sector in Moldova Environmental Financing Strategy*, submitted to the Government of the Republic of Moldova (in English and Russian) prepared by COWI A/S.

DANCEE, OECD/EAP Task Force (2000), *Municipal Water and Wastewater Sector in Georgia Environmental Financing Strategy*, submitted to the Government of Georgia (in English and Russian) prepared by COWI A/S.

DANCEE, OECD/EAP Task Force (2000), Novgorod, *Background Analyses for the Environmental Financing Strategy* (in English and Russian) prepared by COWI A/S.

DANCEE, OECD/EAP Task Force (2000), *Short Justification of the Novgorod Environmental Financing Strategy,* submitted to the Novgorod Oblast Administration (in English and Russian) prepared by COWI A/S.

DANCEE, OECD/EAP Task Force (2001), *Environmental Financing Strategy for Kazakhstan – Background Paper* (in English and Russian), prepared by COWI A/S.

DANCEE, OECD/EAP Task Force (2001), *Environmental Financing Strategy for Georgia – Background Paper* (in English and Georgian) prepared by COWI A/S.

DANCEE, OECD/EAP Task Force (2002), *Environmental Financing Strategy for Kazakhstan – Short justification* (in English and Russian) prepared by COWI A/S.

DANCEE, OECD/EAP Task Force (2002), *Environmental Financing Strategy for the Pskov Oblast of the Russian Federation – Background Paper* (in English and Russian) prepared by COWI A/S.

DANCEE, OECD/EAP Task Force (2002), *Short Justification for the Municipal Water and Wastewater Financing Strategy, Pskov* (in English and Russian) prepared by COWI A/S.

DANCEE, OECD/EAP Task Force (2003), draft report, *Environmental Financing Strategy for the Municipal Water and Wastewater Sectors in the Ukraine, Background Analysis* (in English and Russian) prepared by COWI A/S.

Edwards, C. (2008), "Social Cost-Benefit Analysis – The Available Evidence on Drinking Water", Valuing Water-Valuing Well-Being: A Guide to Understanding the Costs and Benefits of Water Interventions, World Health Organization, Geneva.

European Commission, TACIS in co-operation with the OECD/EAP Task Force Secretariat (2003), *Financing Strategy for Urban Water Supply and Waste Water Treatment in Rostov Oblast.*

European Commission, TACIS (2003) *Development of the Pilot Financing Strategy for Urban Water Supply and Sanitation in Eastern Kazakhstan Oblast.*

Ebinger, J.(2006), *Measuring Financial Performance in Infrastructure: An Application to Europe and Central Asia*, Policy Research Working Paper 3992, The World Bank, Washington, DC.

European Union Water Initiative – Med component (2009), *MED EUWI Egypt Country Dialogue on Water Brief policy document outlining the Dialogue's key findings: Financing water supply and sanitation in the Greater Cairo area,* a document supported by the EUWI-Med component, OECD and the Global Water Partnership, April 2009.

Foster, V. and C. Briceno-Garmendia (2010), *Africa's Infrastructure: A Time for Transformation, African Infrastructure Country Diagnostic Program*, The World Bank, Washington, DC.

Fonseca, C. and R. Cardone (2005), *Analysis of cost estimates and funding available for achieving the MDG targets for water and sanitation*, Leicestershire, Water, Engineering, Development Centre, Loughborough

University; London, London School of Hygiene and Tropical Medicine; Delft, IRC International Water and Sanitation Centre.

Gassner, K., A. Popov, and N. Pushak (2008), *An Assessment of Private Sector Participation in Electricity and Water and Sanitation Services in Developing and Transition Countries*, The World Bank, Washington, DC.

Gasson, C. (2008), "Water and the credit crunch", Volume 9, Issue 10, *Global Water Intelligence.*

Ghosh Banerjee, S. and E. Morella (2010), *Africa's Water and Sanitation Infrastructure: Access, Affordability and Alternatives*, Directions in Development Series, The World Bank, Washington, DC.

Haller L., G. Hutton and J. Bartram (2007), "Estimating the Costs and Health Benefits of Water and Sanitation Improvements at Global Level", Journal of Water and Health, 05.4.2007, WHO, Geneva.

Hutton, G., L. Haller (2004), *Evaluation of the Costs and Benefits of Water and Sanitation Improvements at the Global Level*, Water, Sanitation and Health, Protection of the Human Environment 2004, World Health Organization, Geneva.

Hutton, G. and J. Bartram (2008), "Global costs of attaining the Millennium Development Goal for water supply and sanitation", *Bulletin of the World Health Organisation*, Geneva.

Hutton G, et al. (2008), *Economic impacts of sanitation in Southeast Asia, A four-country study conducted in Cambodia, Indonesia, the Philippines and Vietnam under the Economics of Sanitation Initiative (ESI)*, Water and Sanitation Program, February 2008, Jakarta.

JMP (2010), *Progress on Sanitation and Drinking Water, 2010 Update*, World Health Organization and UNICEF.

Kariuki, M. and J. Schwartz (2005), *Small-Scale Service Providers of Water Supply and Electricity*, The World Bank Policy Research Paper 3727, The World Bank, Washington, DC.

Kingdom, B. and V. Jagannathan (2001), *Utility Benchmarking*, Viewpoint no. 229, The World Bank, Washington, DC.

Komives, K., et al. (2005), Water, *Electricity, and the Poor – Who Benefits from Utility Subsidies?,* The World Bank, Washington, DC.

Lloyd Owen, D (2009), *Tapping liquidity: financing water and wastewater to 2029*, a report for PFI market intelligence, Thomson Reuters, London.

OECD (2003), Social *Issues in the Provision and Pricing of Water Services*, OECD Publishing, Paris.

OECD (2006a), *Infrastructure to 2030: Telecom, Land, Transport, Water and Electricity*, OECD Publishing, Paris.

OECD (2006b), *Keeping Water Safe to Drink*, OECD Policy Brief, March 2006, OECD Publishing, Paris.

OECD (2007), *Infrastructure to 2030, volume 2, Mapping policy for water, electricity and transport*, Paris.

OECD (2008), *Private Sector Participation in Water and Sanitation Infrastructure*, Draft Paper for the Global Forum on International Investment, 27-28 March 2008.

OECD (2009a), *Managing Water for All: An OECD Perspective on Pricing and Financing*, OECD Publishing, Paris.

OECD (2009b), *Strategic Financial Planning for Water Supply and Sanitation, A Report from the OECD Task Team on Sustainable Financing to Ensure Affordable Access to Water Supply and Sanitation*, OECD, Paris.

OECD (2009c), *Private Sector Participation in Water Infrastructure, OECD Checklist for Public Action*, OECD Publishing, Paris.

OECD (2010a), *Pricing water resources and water and sanitation services*, OECD Publishing, Paris.

OECD (2010b), *Innovative Finance Mechanisms for the Water Sector*, OECD Publishing, Paris.

OECD (2011), *Benefits of Investing in Water and Sanitation: an OECD Perspective*, OECD Publishing, Paris.

OECD (2012 forthcoming), *Financing Water Resources Management*, OECD Publishing, Paris.

OECD-DAC (2009), *Measuring aid to water and sanitation*, December 2009, OECD Publishing, Paris.

OECD-DAC (2010), *Financing Water and Sanitation in Developing Countries: the Contribution of External Aid*, OECD Publishing, Paris.

OECD/EAP Task Force, Ministry of Finance and Economy of the Republic of Armenia, (2004), *Financial Strategy for Urban Wastewater Collection and Treatment Infrastructure in the Republic of Armenia* (in English and Russian) prepared by COWI Moscow.

OECD/EAP Task Force (2006), *Financing Water Supply and Sanitation in Eastern Europe, Caucasus and Central Asia*, OECD.

OECD/EAP Task Force (2007), *Implementation of a National Finance Strategy for the Water Supply and Sanitation Sector in Armenia, Assisting the Armenian Water Authorities to Make the Best Use of Available Resources*, OECD, Paris.

OECD/EAP Task Force in co-operation with the European Commission (2008), *National Policy Dialogue on Financing Rural Water Supply and Sanitation in Armenia*.

OECD/EAP Task Force (2007), *Financing Water Supply and Sanitation Sector in Moldova: Executive Report*.

OECD/EAP Task Force (2008), *Facilitating policy dialogue, and developing a National Financing Strategy for Urban and Rural Water Supply and Sanitation in Moldova*, Presentation Report, 2008.

OECD/World Water Council (2008), *Creditor Reporting System on Aid Activities 2008: Aid Activities in Support of Water Supply and Sanitation*.

Marin, P. (2009), *Public-Private Partnerships for Urban Water Utilities, A Review of Experiences in Developing Countries*, PPIAF Trends and Policy Options No 8, Washington, DC.

Martoussevitch, R. (2008), *Regional disparities in the utility sector services in Russia – does the reform of local self-governance help reduce them?*, a report prepared for Public Services International Research Unit, University of Greenwich, June 2008.

Mehta, M. (2008), *Assessing Microfinance for Water and Sanitation: Exploring Opportunities for Sustainable Scaling-Up*, a report to the Bill and Melinda Gates Foundation, Seattle.

Palaniappan, M., et al. (2007), "Water Infrastructure and Water-related Services: Trends and Challenges Affecting Future Development", in *Infrastructure to 2030 (Volume 2), Mapping Policy for Electricity, Water and Transport*, OECD, Paris.

Pinsent Masons (2009), *Pinsent Masons Water Yearbook 2009-2010*, 11th edition, Pinsent Masons, London.

Prüss-Üstün A, et al. (2008), *Safer Water, Better Health: Costs, Benefits and Sustainability of Interventions to Protect and Promote Health*, World Health Organization, Geneva, 2008.

Smets, H. (2008), *De l'eau potable a un prix abordable: la pratique des Etats*, Académie de l'Eau, Paris.

Renwick, M., P. Subedi, and G. Hutton (2007), *Biogas for Better Life: An African Initiative, A Cost-Benefit Analysis of National and Regional*

Integrated Biogas and Sanitation Programs in Sub-Saharan Africa, Draft Final Report Prepared for the Dutch Ministry of Foreign Affairs, April 2007, Winrock International, Little Rock, USA.

Trémolet, S. and D. Binder (2009), *Body of Knowledge on Infrastructure Regulation, Regulatory Challenges: Frequently Asked Questions,* an electronic publication by PPIAF, The World Bank and the Public Utility Research Centre, University of Florida, *www.regulationbodyofknowledge.org/faq/.*

Trémolet, S. and D. Binder (2010), *The Regulation of Water and Sanitation Services in DCs, Literature review, insights and areas for research,* Agence Française de Développement, A Savoir No1, May 2010.

Trémolet, S. with Koslky, P. and Perez, E. (2010), *Financing On-Site Sanitation for the Poor, A Global Six Country Comparative Review and Analysis,* a report for the Water and Sanitation Programme, Washington, DC.

UN-Water (2010), *Global Annual Assessment of sanitation and drinking water (GLAAS),* WHO, Geneva.

World Health Organisation (2008), *Regional and Global Costs of Attaining the Water Supply and Sanitation Target (Target 10) of the MDGs,* WHO, Geneva.

ORGANISATION FOR ECONOMIC CO-OPERATION AND DEVELOPMENT

The OECD is a unique forum where governments work together to address the economic, social and environmental challenges of globalisation. The OECD is also at the forefront of efforts to understand and to help governments respond to new developments and concerns, such as corporate governance, the information economy and the challenges of an ageing population. The Organisation provides a setting where governments can compare policy experiences, seek answers to common problems, identify good practice and work to co-ordinate domestic and international policies.

The OECD member countries are: Australia, Austria, Belgium, Canada, Chile, the Czech Republic, Denmark, Estonia, Finland, France, Germany, Greece, Hungary, Iceland, Ireland, Israel, Italy, Japan, Korea, Luxembourg, Mexico, the Netherlands, New Zealand, Norway, Poland, Portugal, the Slovak Republic, Slovenia, Spain, Sweden, Switzerland, Turkey, the United Kingdom and the United States. The European Union takes part in the work of the OECD.

OECD Publishing disseminates widely the results of the Organisation's statistics gathering and research on economic, social and environmental issues, as well as the conventions, guidelines and standards agreed by its members.

OECD PUBLISHING, 2, rue André-Pascal, 75775 PARIS CEDEX 16
(97 2011 14 1 P) ISBN 978-92-64-12051-8 – No. 59681 2011-02